Mankind and Deserts 3

*Series Editor*
*Françoise Gaill*

# Mankind and Deserts 3

*Wind in Deserts and Civilizations*

*Edited by*

Fernand Joly
Guilhem Bourrié

iSTE

WILEY

First published 2021 in Great Britain and the United States by ISTE Ltd and John Wiley & Sons, Inc.



ISTE Ltd
27-37 St George's Road
London SW19 4EU
UK

www.iste.co.uk

John Wiley & Sons, Inc.
111 River Street
Hoboken, NJ 07030
USA

www.wiley.com



Library of Congress Control Number: 2021930490

British Library Cataloguing-in-Publication Data
A CIP record for this book is available from the British Library
ISBN 978-1-78630-632-6

MIX
Paper from responsible sources
FSC® C013604

# Contents

# Foreword

## Mankind and deserts

Fernand Joly[1] departed from this world before he was able to complete this book, through which he had hoped to share his experiences of and passion for deserts.

"Yet another book on deserts!" some might think; another book to add to the numerous publications dedicated to these alien and fascinating worlds.

This book, however, is different from earlier books, as can be seen from its title "**Mankind and Deserts**". It is based on the singular relationships that are formed between humans and the world of the desert – relationships that are unique because they can be traced back to the very origins of humanity. Indeed, it is from the arid Horn of Africa (East Africa) that large migrations began and it is along the deserts, if not within the deserts themselves, that we find the major cradles of burgeoning historical civilizations. This inhospitable world is also associated with great spiritual leaders such as Moses, Jesus, Mohammed and the Buddha, while serving as the backdrop for adventurers and empire builders from Alexander the Great to Genghis Khan, or from the Incans to the Conquistadors in the Andes and Mexico.

What is this universe that is so barren and yet so mesmerizing?

1 Professor at the University Paris VII, Denis Diderot, who spent 15 years at the Moroccan Institute of Science in Rabat.

A man in the desert: Fernand Joly, Tademaït, Sahara, Algeria

# Introduction: Aeolian Dynamics and Processes

*I was in a world of stark, pure sand dunes stretching away to infinity on either side in strict geometric forms, somewhat affected by the winds but only inasmuch as the same shapes were re-formed (from that point all is calm).*

Jean Giono, "Noé", in *Œuvres romanesques complètes tome III*, R. Ricatte (ed.), Gallimard, Paris, 1974, p. 839.

The action of the wind is perceived to be so omnipresent and preponderant in arid regions that it is still sometimes considered to be even more emblematic of deserts than dryness. Is this why there are so many errors and false information around wind and its best-known manifestation, the dunes? The uniqueness of wind-formed shapes and wind action is the most likely reason for this fascination, leading to some researchers (sometimes quite influential ones) succumbing to the mirage of the wind and overestimating its actions. As we will see, however, the facts are more complex: striking effects are not necessarily synonymous with effectiveness.

Aeolian processes ("aeolian" comes from *Aeolus*, the Greek god of the wind) are among the most original geomorphological processes in the world. The reason behind this is simple: *the circulation of material through the air partially frees the wind, especially over a medium scale, from topographical*

Introduction written by Yann CALLOT. All photographs within this chapter have been provided by the author, unless indicated otherwise.

*constraints* (Figure I.1). The wind thus enjoys **relative freedom vis-à-vis gravity and, therefore, vis-à-vis pre-existing slopes and relief**. For example, along with glaciers, it is the only entity capable of moving material upwards, as certain developers have learned the hard way!

**Figure I.1.** *A sandy spit that curves around relief in the SE of the Tibesti range, in a region with almost unidirectional wind (image from Google Earth). For a color version of this figure, see www.iste.co.uk/bourrie/deserts3.zip*

Given the unique feature of wind actions, we will proceed as follows: we will first more closely examine the conditions and processes that make it possible to form and, sometimes, remake wind formations or aeolian landforms; we will then take a differentiated approach based on the environment and associated formations. Finally, we will conclude with a few representative examples of the interactions (often disastrous!) between the winds and humans.

## I.1. Conditions and environments in which wind acts

### I.1.1. *Where is the wind effective?*

The wind is undeniably unique: it blows across the entire surface of the earth, with varying intensity that is dependent on a number of criteria. These differences can be observed on a global scale. For example, the vast stretches of the Southern oceans are swept by atmospheric depressions that are specific to certain Southern latitudes: the Roaring Forties and Furious Fifties, which are much sought after in sailing races. On the continental level, there are calmer regions, such as central Siberia. The great Canadian north, which is less cold than Siberia, is sometimes considered to have a more difficult winter as it has more violent winds. There may also be marked differences at a local level. A remarkable example of this is the rhodanian corridor: when the northern *mistral* blows, the Rhone valley, oriented N-S, or the plateaus that ring it, may be battered by an intolerable *mistral*, while tributary valleys (such as those with rivers that descend from the Cévennes), which are oriented W-E, experience an almost summery micro-climate.

The omnipresent nature of the wind may lead us to conclude that aeolian landforms exist on all continents. However, this is not the case in reality except over very small areas. Indeed, a few conditions must be met in order for wind dynamics to be effective:

1) Strong and frequent winds are a favorable condition, but not essential. If all other conditions are satisfied, even weak winds can create large aeolian landforms. However, there is a threshold velocity for aeolian erosion below which the wind cannot displace any material. The displacement threshold for sandy materials can vary according to local conditions and, notably, depending on the size of the material being transported. However, it most often lies between 4.5 and 6 m/s, at a height of 10 m. This will be examined in greater detail when we study the quantities of sand that can be displaced by the wind.

2) However, this first condition is not sufficient. In many regions, the wind by itself is not very effective on firm land except during exceptional storms, such as the storms that hit France in December 1999 or Cyclone Xynthia in 2010. It is the substances themselves, which are transported within very precise granulometric limits, that carry out significant morphogenic actions for accumulation and ablation. However, the wind must find available material to transport. Materials that can be picked up and carried are sand, dust (see below), and also volcanic ash, as is sometime seen in Iceland.

3) The previous condition presupposes the existence of at least some portion of bare ground. It thus implies the absence, or at least discontinuity, of plant cover, which usually binds materials and blocks wind action through the roughness and cohesion that it introduces. The reason for this is the strong relationship between green cover and the availability of material. Vegetation causes the soil surface to develop a kind of "roughness" that is so effective at disrupting and preventing displacement by wind that when there is any plant cover over 30%, the wind becomes completely ineffective (Marticorena and

In this last case, the limiting factors may have two origins:

– cold glacial and periglacial climates. The formations could be inherited, like European loess, or present-day formations, further north, in areas where plant cover does not protect the soil;

– arid and semi-arid environments in hot or temperate regions, which are the subject of study in this book, with very little precipitation and high evaporation. Such regions are found in several parts of the world (Figure I.3): Australia, with its interior deserts, the North American West and the arid diagonal through South America; the western part of southern Africa and, above all, the vast arid arc from the Atlantic Sahara to West Africa.

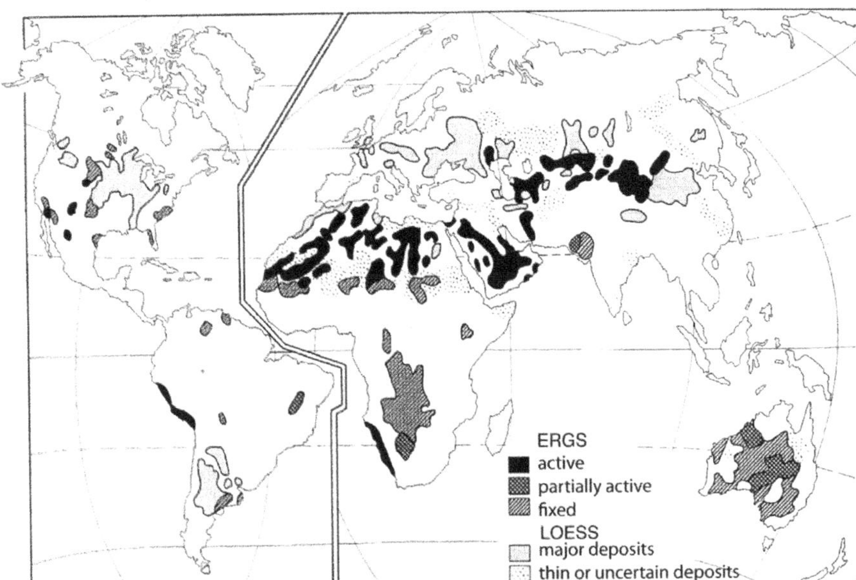


**Figure I.3.** *Distribution of areas with wind deposits around the globe (as per Thomas 1997, modified). Some areas are exaggerated so they are visible on the scale of this map*

This arc covers the whole of the Sahara and its extensions into East Africa. This then continues, interrupted only by the Red Sea and Persian Gulf, into the deserts of Asia, from the Arabian peninsula to the Taklamakan and Gobi deserts, hot in the summer and freezing in the winter. This arc even has some European extensions, such as the Iberian Peninsula.

By listing these "physical" causes, it can be seen that the overall climatic environment, and not just the wind, govern the existence and the efficacy of aeolian action. We also see that these "causes" are not mutually exclusive and can combine, as happens at Cape Juby (see below) where the desert environment and abundant littoral contribution come together to create one of the best and most effective aeolien systems in the world.

#### I.1.2.2. *Anthropized environments*

Since anthropization affects all physical environments, it must be included here as it plays an essential role in arid borders. A rearrangement, in the broadest sense, of soft ground, especially soils, by humans often leads to the more or less complete disappearance of green cover. While this can be observed over large work sites, agricultural activity is the leading cause for this denudation. Occurring all the more easily when the environment is dry, the bare, exposed soils are an important aggravating or triggering factor along arid borders. A catastrophic example is that of the *dust bowl*, an ecological disaster that affected the Southern Plains regions in the United States from 1930 to 1938 (see section 2.3).

## I.2. Aeolian processes

When these conditions come together, the aeolian dynamic depends on several variables as follows:

– the energy supplied by the wind, which is a function of its frequency and its velocity. Turbulence also plays a role, but in conditions that are harder to measure on the ground;

– the direction of the winds. If they blow in all directions, they tend to rearrange the sand without any large migration of material. A similar quantity of energy, in a wind regime that is called unidirectional, i.e., with one marked, dominant direction, results in mobile dune forms (see below) and a large migration of sand that often poses a problem for human activity.

Three kinds of action play a role in wind dynamics: transport, ablation and accumulation. The order in which we study these actions poses a problem as they are in constant interaction. They will be studied here in the same order as they were listed above, as the chief process is transport. However, this order can be debated since ablation, which is how the wind is effective, could also be studied first.

### I.2.1. *Aeolian transport*

The wind transports material in different ways depending on its velocity, and the granulometry, shape, and density of the elements. The sandy material is chiefly made up of quartz, with a few exceptions such as gypseous sands. The process has a few specific features as follows:

– the transport capacity is limited to a rather small granulometric fraction, which will be specified when we examine the types of transport;

– quartz grains called *round matt* (Figure I.4): when examined under a microscope, these have a rounded and blunted aspect resulting from shocks and rubbing against other grains. This makes it possible to distinguish them from rounded shiny grains, which are much more lustrous and are a result of water abrasion from transport through water. The appearance of the round matt grains is due to the repeated shocks to which the grain is subject to during its transport and this produces two effects:

- the surface of the grains is literally *frosted*,

- particularly violent shocks may cause part of the grains to break off and leave behind remarkable, very significant V-shaped traces.

The appearance of the grains is so unique that wind-blown grains can be easily recognized and it is even sometimes possible to see the traces of their journey through the air after the grains have undergone aquatic transport. This is especially true for larger grains, with the smaller grains ($< 250\ \mu m$) rarely being round.

Flat debris is more easily displaced, as it offers the wind a larger surface with respect to its volume. This is the case with seashells on beaches or in paleolakes, or gypsum crystals found around evaporite deposits. The latter are either incorporated into the dunes in the form of thin plates and larger in size than the dunal sands, or form purely gypseous dunes with coarse elements, because of the low density of the grains, related to the high proportion of voids. The famous *White Sands* site in New Mexico is a striking example of these gypseous formations that result in almost completely white dunes that could be mistaken for snow.

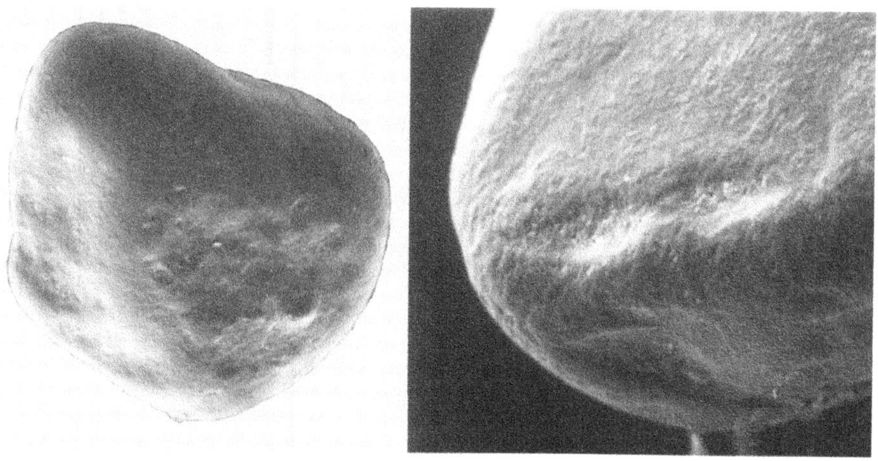

**Figure I.4.** *SEM image of a wind-blown sand grain (in the Taghit region of Algeria), with a diameter of around* 250 µm. *On the right is a close up of a grain with highly blunted traces of aeolian shocks*

Like aquatic transport, aeolian transport is complex: several modes of transport exist and even often co-exist, especially during strong, sandy winds, depending on the size of the elements being transported. These have been extensively studied in laboratory wind tunnels, where wind speed and the quantities displaced can be controlled, and even in the field, with mobile wind tunnels placed along the surfaces of deserts. On simplifying, we can distinguish three kinds of transport based on the size of the element being transported and the wind speed (Figure I.5).

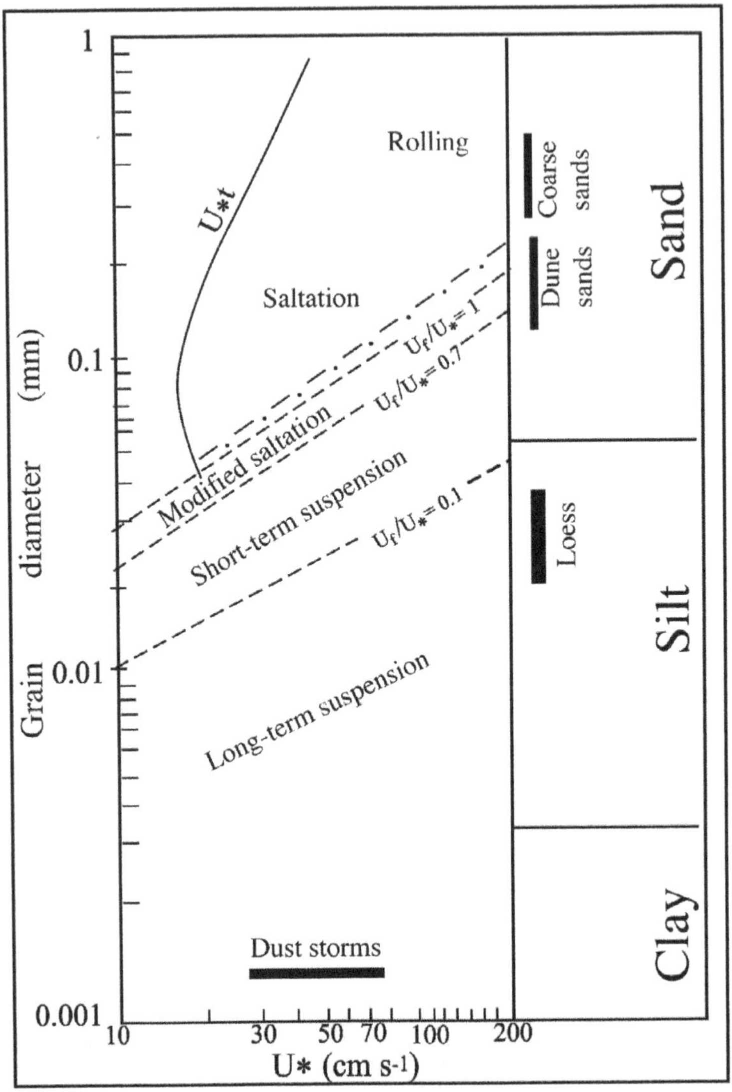


**Figure I.5.** *Relation between the ablation velocity, fall velocity and granulometry in aeolian sediments (according to Tsoar and Pye 1987). Uf is the fall velocity of the elements, U∗ is the suspension velocity and U ∗ t is the threshold velocity for soil to be blown away, and beyond this velocity, grains are lifted up from the soil. It is seen that for small elements (<10 µm), the suspension velocity is so low that the grains practically never fall back to the ground. Rolling occurs above the saltation lines, around the height of the class of coarse sands*

#### I.2.1.1. *Suspension*

A suspension takes up the smallest particles. There are two modes, as follows:

– *A continuous suspension* holds particles with a dimension smaller than 10 µm in the air for long durations. This continuous phenomenon results from the fact that while it is impossible for very small elements to be torn away by the wind (Figure I.5), the wind speed required to maintain them in suspension decreases with their size, such that it is almost zero for a size of around 10 µm. Thus, in theory, particles that are smaller than this must always remain in the atmosphere. Other processes then combine with gravity to get them to deposit. The most frequent cause of this is that these particles play the role of *condensation kernels*. When they arrive in more humid regions, the air in which they are held is loaded with water vapor until it reaches condensation point. However, perfectly clear air does not allow condensation: the presence of very small impurities in the atmosphere (being part of aerosols) is essential as these will serve as condensation kernels. The water vapor then fixes to the particles, changing to liquid state (or solid, in the form of ice, if the temperature is low enough). The drops or flakes that are thus formed continue to grow through the fixation of ambient vapor and then form clouds. If the condensation is great enough, the drops or crystals continue to swell until they reach a size that is too large to allow them to remain suspended in air. At this point, the clouds then produce precipitation, which return particles to the soil that would have been too fine to fall by themselves. *Desert aerosols* (the precise term used for this "dust" carried into the atmosphere) are made up of fine particles, rarely seen as free particles in soils where they are usually in the form of aggregates. They are then liberated by the shocks produced in these aggregated large grains in a state of saltation, when they have enough energy to break them. These fine particles are among the largest sources of solid aerosols in the atmosphere and cover considerable distances once airborne. It has been shown that desert aerosols originating in the Sahara have reached as far as the Amazon (Swap, Garstang, and Greco 1992) and have even contributed to soil formation in the Caribbean (Muhs, Bush, and Stewart 1990). This suspension of fine desert material is the cause for the "blood rains" that sometimes leave ochre or beige stains on cars or drying laundry. In Europe, these "blood rains" are largely composed of Saharan dust and have been referenced in the historical records of certain periods, such as in Antiquity, because they used to be considered ill omens. These rains are an important source of nutrients in the oceans, where they supply the basic minerals (notably iron and phosphorous) for the lowest trophic level (Bergametti et al. 1992; Jickells et al. 2005), just as they contribute to the fertilization of the soil upon which they fall.

– *Discontinuous suspension* affects particles whose size is greater than 10 µm and up to about 125 µm. In these cases, the particles are lifted and held suspended in the air. The finer they are and the more violent the wind, the longer they remain suspended. Particles smaller than 64 µm, or, in exceptional cases, smaller than 100 µm, may be transported over large distances. This has been evidenced by the discovery of windblown sediments in Fuerteventura in the Canary Islands whose provenance is clearly not local (Coudé-Gaussen 1991); as the archipelago is entirely of volcanic origin, its rocks do not contain the material seen in the windblown deposits. Some of the quartz grains in these deposits can be as large as 100 µm. They were, thus, transported by the wind over the 100 km of ocean that separates the closest of the Canary Islands from the Atlantic Sahara to the east. Recent results (van der Does et al. 2018) show that large aeolian particles (>200 µm and till >450 µm) were collected at the sea level in the Atlantic ocean as far as 3,400 km from the west coast of Africa. This cannot be explained by the transfer processes classically admitted (Tsoar and Pye 1987). According to the authors, four processes may combine to explain this observation: very strong winds; turbulence uplifting particles at very high altitudes; negative electric charge carried by some quartz particles due to frictions between them (tribo-electricity) leading to their uplifting in an atmosphere that is positively charged at high altitude; storms and tropical cyclones, favored by the larger thickness of troposphere in the intertropical latitudes.

This downpour of elements in a discontinuous suspension can, by definition, occur without any condensation weighing down the particles and result from a decrease in wind and/or the cessation of the ascending whirlwind-like motion, or the squall-line that lifted them from the soil. A dust fallout, *stricto sensu*, then occurs, with proper clouds of solid material making it as dark as night in the middle of the day. In order to photograph one such dust fallout in Aleg (Mauritania), we had to use an aperture and shutter speed that indicated that about 1/4,000th of the light was able to penetrate through the dust cloud. It was so dark that the birds thought night had fallen and flew back to their nests and fell asleep.

I.2.1.2. *Saltation*

Saltation is the most important process in aeolian transport. The term comes from the Latin *saltare* (to leap) and is a perfect description of the process, which chiefly affects grains of between 100 and 250 μm, or even 350 μm in high winds. There are two factors that contribute to the grains being lifted up from the soil: the wind itself, once it exceeds a certain lifting threshold and the impact of another grain falling on the surface, which transmits its energy to the grain in the soil, like a billiard ball in motion striking a stationary ball. The grain that is lifted up performs a short "hopping motion", which varies depending on the nature of the substrate (Figure I.7): on a pebbly surface, the grains rise upwards almost vertically, before re-descending at a slight, nearly constant angle, covering a few decimeters or meters.

**Figure I.6.** *Dust falling over Aleg (Mauritania) in April 2001. Colors not modified. Pellicule Ektachrome 100 ASA, 1 sec. at f : 2. For a color version of this figure, see www.iste.co.uk/bourrie/deserts3.zip*

Over a continuous sandy surface, the hops are much shorter in height and length. Thus, when the wind is "normal", the majority of the grains are not lifted very high, traveling over a few decimeters for the most part. When the wind blows with a velocity that is slightly higher than the sand mobilization threshold, there arises the "sand blowing" phenomenon: with saltation occurring through hops or leaps that are only a few centimeters in height, the soil appears fluid while the air remains clear (Figure I.8).

### I .2.1.3 *Reptation*

Reptation displaces larger grains, which cannot be lifted from the surface but retain the ability to roll, even if the etymology of the term, the Latin *reptare* (crawling), suggests another kind of movement. The origin of the displacement is the same as in saltation: the grains roll under the wind's influence, but their displacement is often initiated by the impact of smaller grains. Thus, there is both rolling as well as a bombardment effect, and, consequently, this phenomenon is only seen during more violent winds.

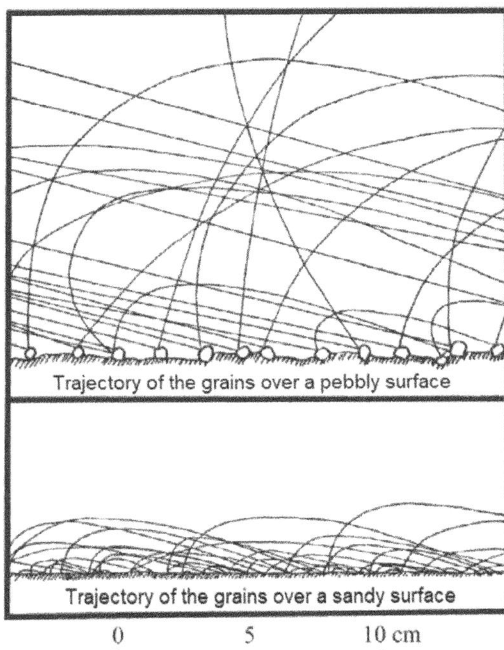


**Figure I .7** *How the roughness of the substrate influences saltation (as per Bagnold 1960). Over a pebbly substrate, the grains perform large hops; on a sandy substrate, these hops are shorter*

The maximum size of the grains that can be displaced through reptation is the subject of much debate. Along the edge of the large Saharan ergs, grains larger than 2,000 μ m (2 mm) are picked up by the wind and form large ridges (see below) that can reach a height of 30 cm. In favorable topographic conditions, with flat ground, wind displacement can even affect grains that are 5,000 μ m (5 mm) in size (Barbier 1987).

The dividing lines between the different types of transport are not as clearly drawn as this classification would suggest. Like all classifications, it is simplistic and does not reflect the continuum of geomorphological forms and processes: first of all, suspension, saltation and reptation may all be observed at the same time; second, the whole system may shift to include larger grains depending on the wind force. Saltation and reptation can often occur between 250 and 500 µm, or with even larger grains, during heavy sandstorms. Finally, turbulence may modify any of these processes. Some authors (Anderson 1987) identify a modified saltation where the leaping of the grains is disrupted by turbulence.

**Figure I.8.** *Blowing sand in the Tassili of Hoggar, between Tamanrasset and In Guezzam. For a color version of this figure, see www.iste.co.uk/bourrie/deserts3.zip*

I.2.1.4. *Quantification*

It is important for the people in charge of constructing installations around the desert to estimate what quantity of sand is transported by the wind, as they are often confronted with unexpected sand encroachment. This quantification is also necessary to understand the creation of aeolian accumulation forms, such as sand dunes. However, the quantities displaced are often more difficult to study than the modes of transport, given the difficulties around carrying out measurements on the ground and the high number of variables involved. The measurements themselves are complicated by the very presence of the measuring devices, especially traps. The obstacle the trap creates can itself disrupt displacement to such an extent that sometimes it is completely unreliable. However, simple, robust and efficient devices have been developed to measure saltation flux (Fryrear 1986). Physicists and physical chemists who study the atmosphere have also managed to obtain reliable readings of the flux of particles that can be transported over long distances. They use measuring towers, with anemometers and filter holders at different heights that make it possible to estimate the cumulative quantity of fine material passing over a given substrate.

Another approach, inspired notably from the work of Fryberger (1979), seeks to resolve this problem by measuring displacement using an approach that studies energy: replacing the measurement of the phenomenon by a measurement of its cause, the wind. The reason for this is simple: a series of measurements of wind velocity and the direction of the wind already exist in a number of stations in the desert. Although the number of stations is sometimes quite small (as these are arid and poorly populated regions), the data are still sufficient to allow comparisons to be made between deserts and to be verified with comparisons to aeolian landforms situated around these stations. The transport formulae, developed for the wind tunnels, are quite difficult to apply to measurements on the ground and so Fryberger has proposed a more pragmatic formula, which is simplified to allow it to be used with wind data alone. The formula no longer seeks to estimate the quantity of sand transported, but the potential for the sand to be lifted, $q$, which is calculated based on the wind data alone and, therefore, based on available wind energy.

Fryberger believes that threshold wind velocity for aeolian erosion is proportional to the wind velocity at a given height, a relationship that has been demonstrated by Bagnold (1960). In the first approximation, he replaces the aeolian erosion threshold by the wind velocity at 10 m (the standard height for anemometers), obtaining:

$$q \propto V^2(V - Vt), \qquad [I.1]$$

where $\propto$ expresses a proportionality ratio, and not an equality, $V$ is the wind velocity (in knots) at a height of 10 m, $V_t$ is the impact threshold of the wind velocity at 10 m (minimum velocity in knots at 10 m required to maintain the sand in a state of saltation; he fixes this at between 5 and 6 m/s for sand in active dunes).

Formula [I.1] shows that the increase in the potential sand transported is proportional to the velocity cubed from the moment the transport begins. This makes it possible to calculate a sand transport coefficient for each direction the wind rose, obtained by multiplying the annual data for each speed by the coefficient that corresponds to the velocity of the measurement class ($x$ m/s in general). In order to simplify the calculation, conventionally the transport coefficients obtained using formula [I.1] are divided by 100 for measurements in knots (Fryberger 1979) and by 12.5 for measurements in m/s (Callot 1987; Callot and Oulerhi 1996).

Comparisons with measurements and observations on the ground have shown that the simplifications carried out conserved the validity of the measurements when all transport conditions came together. For example, in the case of the "sand river" in Cape Juby (Atlantic Sahara), studied by Oulerhi (1992) whose maritime origin allows for a limitless supply of sand (see the later section on deflation). Thus, through a digital processing of the information on the wind, it is possible to calculate the potential transport energy in a given region and construct "sand mobilization roses" (like wind roses) of different types (Fryberger 1979), and to calculate the result of these roses. A conversion formula can then allow us to move from the coefficients to the quantities of sand transported. In the very windy regions of the sand river in Cape Juby, close to Tarfaya (Morocco), the calculation yields a migration figure of around $200\,\mathrm{t\,m^{-1}\,yr^{-1}}$. This means that in a section of 1 m, perpendicular to the resulting winds, 200 tons of sand transit through every year in the direction of the resultant. This figure has been verified by measurements on the ground made by teams using equipment to remove soil from the roads (for more details on the techniques and computer processing, see Callot and Oulerhi (1996)).

### I.2.2. *Ablation*

This is a purely erosive action (in the strictest sense of the term) and is often "discreet" in its effects on the landscape. However, it is essential to study this phenomena, especially for understanding the interactions between the physical environment and the societies that live there. This phenomenon takes place in two forms, which may both combine at the same spot.

#### I.2.2.1. *Deflation*

Deflation is the mobilization of material by wind – this material can either be loose or have come loose from some initially coherent material that was altered through other processes (hydrological alteration, for example). As wind can only carry away materials within its transport capacity, which is limited to elements of specific sizes (Figure I.5), deflation is a process that involves selection, sometimes called *aeolian sorting*, and results in a deflation pavement (a result attributed to deflation, at any rate), which is a thin layer made up of material whose granulometry is higher than that which the wind can transport (Figure I.9).

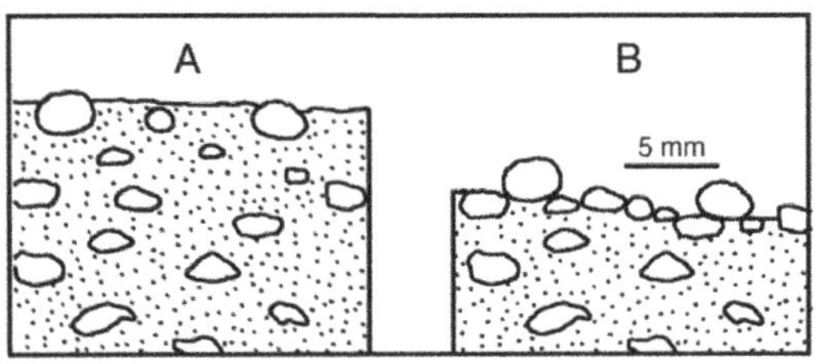


**Figure I.9.** *The mechanism of formation of a deflation pavement: (A) the initial state; (B) the wind carries away fine material from among the heavier materials, which get concentrated on the surface until they adequately cover the finer elements and protect them from wind ablation*

It must be noted that this granulometry does not necessarily correspond to elementary granulometry: for example, the aggregates frequently seen in clayey material (*micropeds*) have a short-term "aeolian" behavior (they may then be destroyed) corresponding to the overall size of the aggregate.

A clear example of the deflation pavement effect is that of wheel tracks. Ever since the automobile was invented, there have been many accounts of travels through deserts that mention the adventures (sometimes with a tragic end) of people who got hopelessly lost by following tire tracks that led nowhere because they were, in fact, very old. When a vehicle passes through a region where there are quite loose soils, it leaves an impression in the soil by destroying the deflation pavement. The wind then carries out the same "winnowing" process in the tire tracks and creates a fresh deflation pavement within the tracks that "fixes" them. If the region is very arid, the rare precipitation might slowly flatten the trace, or a light layer of soil might "fossilize" it. These tracks can take decades to disappear, especially in a region that is hyper-arid.

The deflation pavement is so effective that it is able to preserve superficial formations of astonishing fragility. One of the most remarkable examples of this is the *fech fech* (Figure I.10). The fech fechs – the plural form is used here as they have diverse origins – are pulverulent formations with no cohesion, very low density and sometimes resembles flour. Exposed to the wind, or thrown up by a passing car that drove through it, they are capable of rising tens of meters into the air. But these formations are stable and are preserved over millennia, at least, given that some of those seen today are lacustrine deposits from the Holocene (Callot 1987), protected only by the formation of a deflation pavement – a precarious but effective protection.

**Figure I.10.** *A vehicle that has "fallen" into a fech fech outcrop of the Mekheranne sabkha (Central Sahara, Algeria). In the foreground, the deflation pavement contrasts starkly with the pulverulent substratum thrown up by the wheel. For a color version of this figure, see www.iste.co.uk/bourrie/deserts3.zip*

If the pavement is regularly destroyed by a natural or anthropic action, there may be a continuous supply of transportable material. This "wellspring" for the "river of sand" in Cape Juby, close to Tarfaya (Morocco (Figures I.11 and I.12)) is an excellent example of this: along this coast you have the remarkably constant trade winds, blowing NNE-SSW, which approach, at an angle of about 60°, a coastline along which there is a ENE-WSW littoral drift that transports sand in abundant quantities.

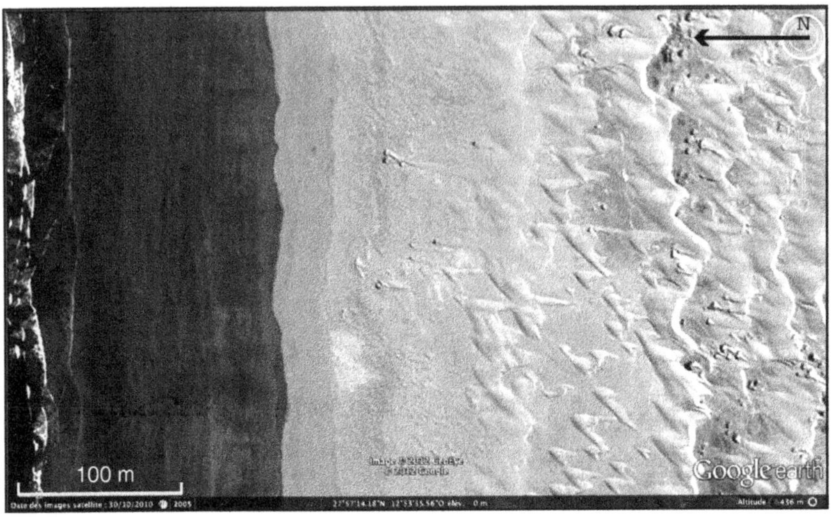


**Figure I.11.** *Satellite image of the area where the "river of sand" begins in Cape Juby, close to Tarfaya (image Google Earth on October 30, 2010). North lies to the left. In the north, we can see a zone with waves and then 100 m of dark, damp sand. At* 100 m *from the beach you have the first dune accumulations*

Further, the tidal range can be as much as 3 m, and produces vast, sandy beaches. The wind dries the sand and carries it inland (Figure I.11). There is a very clear line marking the passage from wet sand to dry sand (Figures I.11 and I.12(a)). The earliest dunes appear at around 100 m from the shore (Figure I.12(b)), forming "upstream" of the "river of sand" that transforms into a mass of coalescent barchans (Figure I.12(c)). Each high tide destroys the deflation pavement that has started to form on the beach sand and the process is resumed at the next low tide, i.e. twice a day. The problem of silting over, experienced by the nearby Tarfaya port, attests to the efficacy of this alternation of processes.

(a) The Cape Juby beach at low tide. In the foreground is the area of sand dried by the wind, which forms small furrows in the dried sand.

(b) One hundred meters from the high tide line mark you have the first barchanic structures.

(c) A few hundred meters from the shore, a small erg is formed by the coalescence of the barchans.

**Figure I.12.** *Views of the soil in the area where the "sandy river" of Cape Juby begins. For a color version of this figure, see www.iste.co.uk/bourrie/deserts3.zip*

#### I.2.2.2. *Corrasion*

Corrasion (not to be confused with corrosion) is the wearing away of a hard or coherent rock (clay, for example) by material, especially sand, transported by the wind. The process is only effective up to a limited height as the bulk of the transported material is only lifted about 1 m above the soil or, sometimes, only a few centimeters (Anderson and Haf 1988). Often combined with other weathering processes, this generates "mushroom shapes" (see the below section on wind erosion forms). While this process can lead to some spectacular formations, it only plays a minor role in aeolian ablation as the wearing away of coherent material through abrasion only from the material transported by the wind is very difficult except in highly favorable conditions.

### I.2.3. *Accumulation*

Accumulation is, above all, seen in dunes formed by sands and deposits of fine materials, such as loess. The different forms of accumulation will be studied in the section on aeolian landforms. These occupy large areas (Figure I.3), even if we only consider formations that are thick and widespread.

The conditions for this accumulation may arise from:

– a decrease in wind and, therefore, a decrease in transport energy;

– a supply of sand that is so large that the wind no longer has the capacity to transport all the aeolian material and must deposit part of it;

– a significant increase in the roughness of the soil, which leads to the first condition: The energy is dissipated to such an extent by obstacles on the soil that it becomes ineffective at this level, as was seen above when studying conditions for wind action. Vegetation may play this role effectively in peri-desertic regions.

### I.2.4. *The combination of processes*

The combination of processes makes it possible to perceive the capacities and limitations of wind action. While accumulation can clearly be perceived through the large forms it produces (dunes, in the case of sand, or thick layers in the case of dust or loess), it has been seen that it is more difficult to quantify the transport. Ablation is even more difficult to study, given the number of variables that must be monitored for an estimation, whether for sand or for dust (Callot, Marticorena, and Bergametti 2000).

**An essential concept** in the study of aeolian phenomena must be recalled. Although this may seem obvious to some people, it has been vehemently denied by other researchers, carried away by their enthusiasm for wind action: **the wind is a remarkable agent of transport**, even if only for a highly restricted granulometric class of the sizes listed above. **However, it is impossible for the wind to create the material it transports**, apart from the small amount of material created through corrasion. **Thus, it is effective only on loose rocks, which are most often sandy, or on coherent rocks weakened by other agents**. This feature clearly sets apart wind action from processes related to water and dissolved salts, seen in previous volumes. Indeed, water is able to alter whole, coherent rock and then carry away the material produced by alteration, while the wind is only a tributary to other processes that prepare the material for it to act upon. In addition, we pointed out that in deflation a sorting is carried out on the surface that gradually forms a deflation pavement, a type of desert pavement, which, if undisturbed, protects and preserves the surface, as has been demonstrated through the conservation of tire-tracks for decades, in very arid zones.

Because of these limitations, and contrary to what we may think, an alternation of processes is conducive to intense aeolian activity: a non-eolian process creates or supplies material that the wind will then take up. This alternation is largely found in semi-arid and arid regions. In hyper-arid regions, wind activity may be the only practically effective agent, but we may find even this effectiveness highly reduced in landscapes that are almost fixed in place due to a lack of available material. Cape Juby, mentioned several times before, is one of the most remarkable examples of alternating processes, but this is quite a rare occurrence. On the other hand, an identical and frequent example in a semi-arid milieu is that of wadis whose leeward shores are lined by dunes: every flood brings in sandy material, which is picked up by the wind when the wadi dries out. A deflation pavement is formed, which is then destroyed by the following flood. This provides a frequent and renewable supply of sand, especially if the catchment basin contains sandstone outcrops. The Saharan Atlas, in Algeria, which borders the Sahara, contains certain excellent examples of this, such as the Kheneg Menhar which we will examine in later chapters, when we study green barriers (section 2.4, p. 91). In a hyper-arid region, where wadis almost never flow, this system is less effective: wind transport here is almost never fed by fluviatile inflows.

### I.3. References

Alsharhan, A., W. Glennie, G. Whittle, and C. Kendall (1998). *Quaternary Deserts and Climatic Change. Proceedings of the International Conference on Quaternary deserts and climatic change*. Rotterdam: Balkema. XIV+622.

Anderson, R. (1987). "Eolian sediment transport as a stochastic process: the effects of a fluctuating wind on particle trajectories." *Journal of Geology* 95, pp. 497–512.

Anderson, R. and P. Haf (1988). "Simulation of eolian saltation." *Science* 241, pp. 820–823.

Bagnold, R. (1960). *Physics of Blown Sand and Desert Dunes*. 3rd ed. London: Methuen. 265 pp.

Barbier, J. (1987). "Arid environment geochemical exploration in Saudi Arabia - The problem of aeolian contamination." *Journal of Geochemical Exploration* 27, pp. 29–51.

Bergametti, G., E. Remoudaki, R. Losno, E. Steiner, B. Châtenet, and P. Buat-Ménard (1992). "Sources, transport and deposition of atmospheric phosphorus over the Northwestern Mediterranean." *Journal of Atmospheric Chemistry* 14, pp. 501–513.

Callot, Y. (1987). *Géomorphologie et paléoenvironnements de l'Atlas Saharien au Grand Erg Occidental : dynamique éolienne et paléolacs holocènes*. Mémoire des Sciences de la Terre 87-21. Paris: Université Pierre et Marie Curie. 412 pp.

Callot, Y., B. Marticorena, and G. Bergametti (2000). "Geomorphologic approach for modelling the surface features of arid environments in a model of dust emissions: application to the Sahara desert." *Geodinamica Acta* 13 (5), pp. 245–270.

Callot, Y. and T. Oulerhi (1996). "Géodynamique des sables éoliens dans le Nord-Ouest saharien : relations entre aérologie et géomorphologie." *Geodinamica Acta* 9 (1), pp. 1–12.

Coudé-Gaussen, G. (1991). *Les poussières sahariennes - cycle sédimentaire et place dans les environnements et paléoenvironnements désertiques*. Paris: John Libbey Eurotext. 498 pp.

Fryberger, S. (1979). "Dune forms and wind regime." *A Study of Global Sand Seas*. Ed. by E. M. Kee. US Geological Survey Professional Paper 1052. Washington, D.C.: U.S. Government Printing Office, pp. 137–169.

Fryrear, D. (1986). "A field dust sampler." *Journal of Soil and Water Conservation* 41, pp. 117–120.

Glennie, K. (1970). *Desert Sedimentary Environments*. Amsterdam: Elsevier. 222 pp.

Jickells, T. et al. (2005). "Global iron connections: between desert dust, ocean biogeochemistry and climate." *Science* 308 (5708), pp. 67–71.

Marticorena, B. and G. Bergametti (1995). "Modeling the atmospheric dust cycle: 1. Design of a soil-derived dust emission scheme." *Journal of Geophysical Research* 100, pp. 16415–16430.

Muhs, D., C. Bush, and K. Stewart (1990). "Geochemical evidence of Saharan dust parent material for soils developed on Quaternary limestones of Caribbean and western Atlantic islands." *Quaternary Research* 33, pp. 157–177.

Oulerhi, T. (1992). *Étude géodynamique des migrations de sable éolien dans la région de Laayoune (Nord du Sahara marocain)*. Mémoires des Sciences de la Terre 92-12. Université Pierre et Marie Curie (Paris 6). 223 pp.

Swap, R., M. Garstang, and S. Greco (1992). "Saharan dust in the Amazon basin." *Tellus* B44, pp. 133–149.

Thomas, D. (1997). *Arid Zone Geomorphology*. London : John Wiley & Sons Ltd. 713 p.

Tsoar, H. and K. Pye (1987). "Dust transport and the question of desert loess formation." *Sedimentology* 34, pp. 289–299.

Van der Does, M., P. Knippertz, P. Zschenderlein, R. Giles Harrison, and J.-B. W. Stuut (2018). "The mysterious long-range transport of giant mineral dust particles." *Science Advances* 4.12, pp. 1–8. doi: 10.1126/sciadv.aau2768. eprint: https://advances.sciencemag.org/content/4/12/eaau2768.full.pdf. url: https://advances.sciencemag.org/content/4/12/eaau2768.

# 1

# Aeolian Landforms in Deserts

## 1.1. Forms and surface states

The wind generates specific forms. Their features are often related to the processes that created them. Starting from the postulate that transport itself does not create forms, we can make a distinction between two types: those related to the action of the transported material (ablation forms) and those produced by the accumulation of this material.

The term "forms" or "landforms" will be used in a broad sense here to resolve one of the major problems encountered in earth sciences: the impossibility of classifying the objects being described into *species*, as in the biological sciences. This implies that the continuum of these objects will be broken up into "forms" (whenever they can be characterized). Particularly in the study of "aeolian processes", the ubiquity of wind action on the earth's surface, and the absence of any hierarchy of the forms through the convergence of elementary objects (as is the case in hydographic networks, organized by gravity flow), leads to the study of objects of such varying sizes that not all of them can be categorized into relief forms.

These small objects do, however, play an essential role through their number and how they are organized, often covering vast stretches without any notable thickness, which implies that they cannot be characterized by volume.

---

Chapter written by Yann CALLOT. All photographs within this chapter have been provided by the author, unless indicated otherwise.

This has led to researchers prioritizing the concept of a "surface state", a highly rigorous concept used by researchers studying the physical chemistry of the atmosphere in order to characterize the mobilization capacities of material on the soil surface. This characterization is sometimes based on a geomorphological approach that makes it possible to model the ruggedness of the desert surface. The model that is thus obtained is itself applied to other models for the creation of desert aerosols, greatly improving the quality of the results (Marticorena et al. 1995; Callot, Marticorena, and Bergametti 2000).

This distinction between surface states and landforms makes it possible to be more rigorous in considering the variety of wind effects, while respecting the varying scales across which this is observed: quite often an aeolian landform will be covered by one or more surface states; on the other hand, "aeolian" surface states may exist without being overlaid onto aeolian landforms.

## 1.2. Ablation forms

Ablation forms are much larger due to the vast stretches of surface states that are generated by ablation, rather than the landforms themselves, which actually cover a more limited area on the scale of global deserts.

### 1.2.1. *Desert pavement*

A desert pavement is a surface state that affects a variety of substrates: hard or soft rock, superficial formation.

This feature is sometimes confused with a deflation pavement, but the term must be used with caution as it designates a single, unique cause to the formation of the pavement. While wind is the dominant process at present in arid and semi-arid regions, many other processes can, or could have, contributed to the formation of such pavements or could even have been the principal cause of their formation. These include weathering, which prepares the material for the wind, and fluviatile action, which could have brought in the constituent elements of the pavement and which is often responsible for the flattening of surfaces and the spreading out of material. Observations in the Mojave Desert (Wells, Dohrendwend, et al. 1985) even indicate that certain pavements may have resulted from an accretion process: a pavement of volcanic debris was separated from its original flow by a meter of windblown sediments. The sediments were only able to be deposited by the "infiltration" of these sediments between the constituent elements of the pavements, and then by the coarser sediment rising up through the windblown material, like pebbles and blocks rise to the surface of a field.

(a) Limestone reg 200 km to the west of Luxor (Egypt). The climate is so arid that corrasion predominates and forms networks of grooves in the rocks. On the right: a close-up, the pen in the red circle for scale.

(b) Reg on the piedmont of the Saharan Atlas (Algeria). The slope is enough for run-off to combine with the wind to displace coarse elements. A xerophytic bush and an ephemeral plant manage to survive on this surface that is bare of any fine material apart from a thin sandy cover.

**Figure 1.1.** *Reg. For a color version of this figure, see www.iste.co.uk/bourrie/deserts3.zip*

There are three names for this pavement, depending on the size of the constituent material:

– **the reg** (Figures 1.1(a) and 1.1(b)) is the best-known. It is made up of a series of blocks or pebbles, which are more or less closely arranged, with dimensions varying between a few centimeters and, sometimes, 1 or 2 dm. A reg may have a hydrological origin, especially along a visible slope, but the aeolian dynamic cannot, of course, be excluded. This origin may be verified based on the shape of the constituent elements: if they are rounded, a fluviatile origin may be assumed, unless the substrate is made up of puddingstone. But we must then study the processes that lead to the release of these elements, as fluviatile pebbles may have been released by aeolian processes. On certain surfaces that are almost perfectly flat, like the large hamadas in the Northern Sahara, or the vast Central Saharan plains, the absence of any slope makes it possible to state that the wind plays the dominant role in these formations;

– **a serir** (Figure 1.2) is less well-known and is made up of much smaller constituent elements than a reg, their dimension being smaller than 5 cm. Often rounded by the wind and/or water, these elements are also more or less spaced out. Although it is made up of smaller elements, the serir may also offer remarkable continuity and can look identical for over a thousand square kilometers, without the causes leading to these formations being clearly identified;

**Figure 1.2.** *A serir close to Doush (south of the Kharga depression, Western Desert, Egypt). For a color version of this figure, see www.iste.co.uk/bourrie/deserts3.zip*

– the granulometric continuum of elements covering desert surfaces makes it difficult to define an exact line separating serirs from surfaces made up of large sand grains that have undergone little or no displacement by the wind, and which protect varying underlying surfaces (clays, silts, sands). The stability of these covering layers often lead to an "alteration" of the grains that are ferruginized and become darker. This formation is sometimes called a *rust layer* (Figure 1.3). This also covers vast surfaces, especially in hyper-arid regions where the surfaces are often almost permanently fixed by the absence of any alternation in the processes (see above).

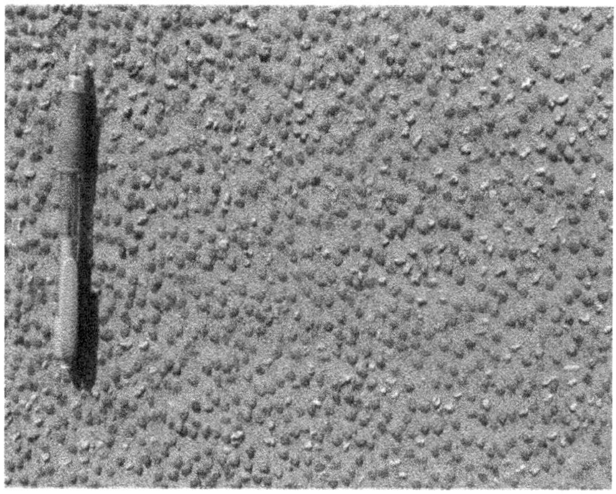

**Figure 1.3.** *Discontinuous rust layer in the Kharga depression (Western Desert, Egypt). For a color version of this figure, see www.iste.co.uk/bourrie/deserts3.zip*

Regs, serirs and rust layers together cover the larger part of desert regions. They cover a much larger area than ergs. In fact, they appear to be the "forgotten ones" within texts and work on desert landscapes, since they are taken so much for granted (Roux 1996). However, they must not be likened to this landscape or associated with it, as we can find regs in mountains, on plains, on plateaus, or even in the spaces between dunes in ergs. They play a significant role in preserving desert surfaces.

### 1.2.2. *Coherent rock landforms*

Coherent rock landforms are relatively underdeveloped, given the capacity of low winds to hollow out hard rock. Remarkable forms may appear in these rocks if the actions are allowed to be carried out over adequate time, however most coherent rocks simply produce soft, often recent sediments, most often as superficial formations.

#### 1.2.2.1. *Minor forms*

When acting on hard, homogeneous rocks, the wind creates ventifacts or *Dreikanters* (Figure 1.4); these pebbles remain in place over long periods (several hundreds or thousands of years) without any change in position. They are thus polished by the wind into aerodynamic forms and present slightly concave facets in the form of three-sided pyramids (hence, the name *Dreikanter*).

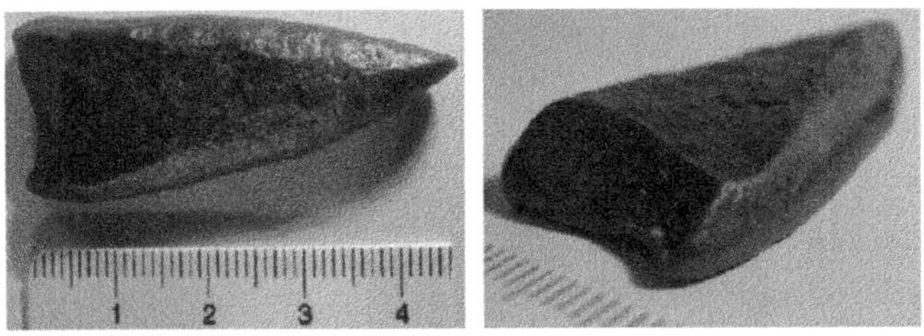


**Figure 1.4.** *Dreikanter in Ténéré (Niger) (photos F.M. Callot). For a color version of this figure, see www.iste.co.uk/bourrie/deserts3.zip*

**Figure 1.5.** *Corrasion lines on a sandstone outcrop in Nubia, in the Kharga depression (Western Desert, Egypt). For a color version of this figure, see www.iste.co.uk/bourrie/deserts3.zip*

The wind may also carry out differential ablation, highlighting irregularities, especially discontinuities in the rock, fractures or stratification joints. It then creates corrasion lines that cause thin furrows in the stationary

They have a perfectly aerodynamic form if the wind has a dominant direction and if the material is homogeneous, with the leeward end being tapered. These are much more frequent than one would think and, like the corrasion furrows, are excellent indicators of the mean annual wind efficacy. To be more precise, they indicate the resultant migration of sand grains over long periods (thousands of years or more). In Algeria, for example, we find them to the north of the Grand Erg Occidental in the southern part of the Hamada de Tinrhert, in the Paleozoic outcrops of the Sebkha Mekkerrhane where they form lines of protuberances dividing the very hard Paleozoic layers of the Appalachian relief (Figure 1.9). The rate at which they are formed (slow in hard rock) may be very fast, particularly in recent times, softer superficial formations like Holocene lacustrine deposits (of the order of 1 mm/yr at least) or desert loess. In the very soft Pleistocene sandstone formations located just north of Nouadhibou (NW of Mauritania), these protuberances can develop to such an extent that they form the majority of the landscape. A similar occurrence is seen, over a smaller area, in the desert loess in the Kharga depression, Egypt (Figure 1.10).

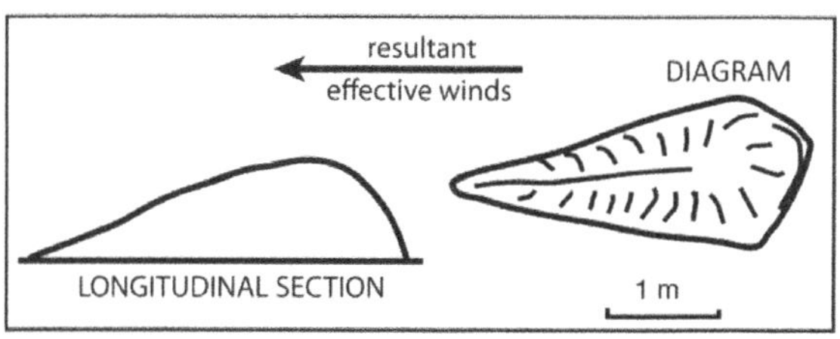


**Figure 1.8.** *Yardang in a homogeneous coherent rock in a region with monodirectional wind (longitudinal section and diagram)*

### 1.2.2.3. *Ridge-Corridor systems*

The most original ablation shapes created by the wind, Ridge-Corridor systems (RCS), can be seen in only a handful of regions (Mainguet 1970; Mainguet 1972). While being quite numerous, they are present in very few areas and the best-known structures are in the sandstone plateaus of Djado (the Algerian-Nigerian border) on the periphery of the Tibesti; and in a different form, in the Lut desert in Iran, where they are called *kalut*. They also exist in the Western Desert of Egypt and in the deserts of China. These structures can be tens of kilometers long, their width can range from hectometers to kilometers and they can be tens of meters in height.

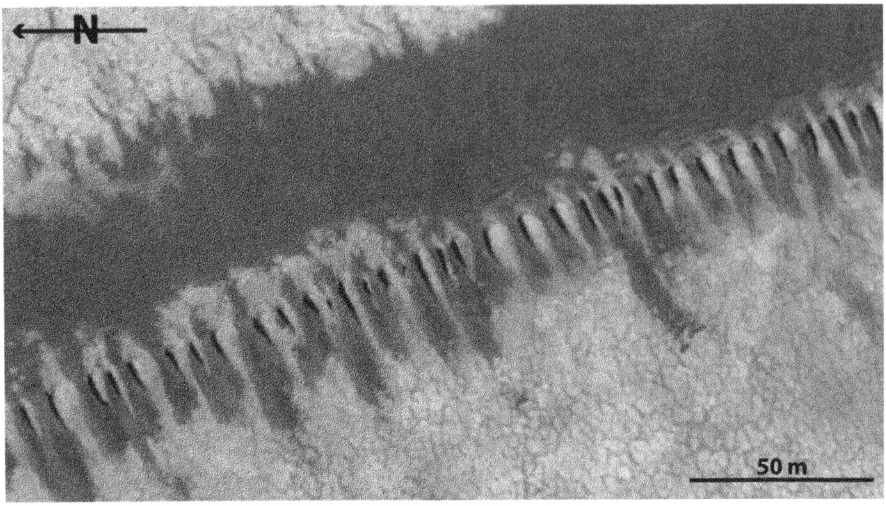


**Figure 1.9.** *Line of yardangs along an Appalachian outcrop of hard rock in the Sebkha Mekkerrhane (south of Akabli, in the central Algerian Sahara). For a color version of this figure, see www.iste.co.uk/bourrie/deserts3.zip*

Attributing the formation of these RCS to wind action alone raises several problems. As noted already, aeolian ablation through corrasion only occurs at low heights, on a decimetric level. Even if we assume that there is a progressive deepening of the furrows separating the ridges, they should present overhangs and develop through collapsing, which does not happen. Their genesis is more complex, as we already saw in the conclusion to the section on processes, it is associated with a combination of processes rather than any one single process. The RCS are initiated by the wind hollowing out fracture joints parallel to or sub-parallel, up to a few degrees, to the prevailing wind. Ridges like this can be seen in Bembéché, SE of Tibesti (Figure 1.11(a)). Although this kind of hollowing out cannot create large shapes, even minimal deepening of these fractures can produce a significant effect when combined with other processes: the small, long depression formed by the wind concentrates flows and allows basic runoff and/or allows water to stagnate within the fracture.

This leads to alteration of the rock and, more generally, to weathering actions that facilitate corrasion and deflation; the deflation pavements created are then destroyed by water flow. Further, water bodies sometimes follow the direction of the winds but are divided by the principal streams.

(a) Destruction, through corrasion, of a soft outcrop of desert loess. The ruins of a Ptolemaic temple can be seen in the background and the vestiges of a blocked-up water channel or well, a *qanat* (or foggara), in the foreground show that the yardangs have been formed since Antiquity, through the ablation of 4 m to 5 m of desert loess over a large part of this area.

(b) Yardangs seen in profile. The wind-facing side is to the right. Roman ruins at the peak of certain protuberances also show that the carving out of these structures is later than Antiquity.

**Figure 1.10.** *Yardangs in the Doush region (Kharga depression, Western Desert, Egypt). For a color version of this figure, see www.iste.co.uk/bourrie/deserts3.zip*

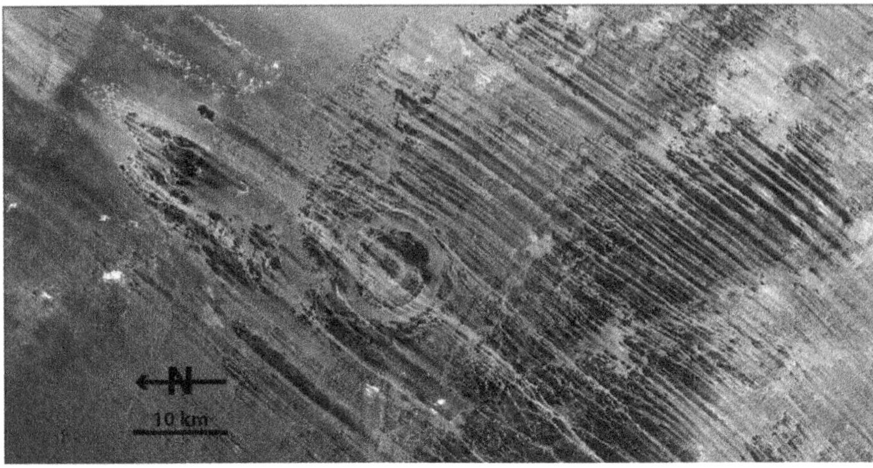


(a) Overview of a dissected plateau. At this scale, it can be seen that the wind can deviate slightly from its path based on the principal directions of the fractures it encounters.

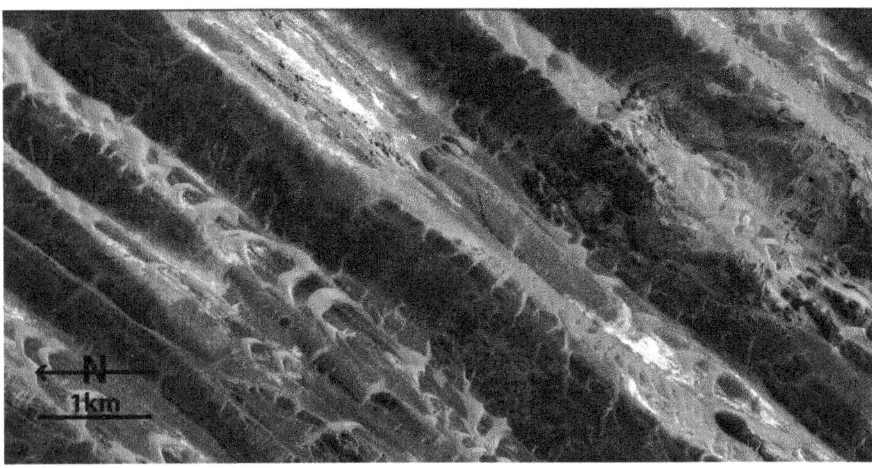


(b) A more detailed view of the central part. Lines of barchans run through the corridors and white sandbanks indicate the existence of water bodies, probably dating to the Holocene. The numerous fluviatile indentations show that water played an important role in the dissection of the ridges.

**Figure 1.11.** *Overview of the Ridge-Corridor system in the SE of the Tibesti (source: Google Earth). For a color version of this figure, see www.iste.co.uk/bourrie/deserts3.zip*

Aeolian exportations also allow juxtaposition, within the same fracture, of sections where water runs off and depressed areas where, in wet periods, water bodies form leaving behind lacustrine Paleolake sediments (Figure 1.11(b)).

When examined in more detail, it can also be observed that the "ridges" are subject to the action of local run-off, which sometimes indents the sides of the relief. Seen as being regularly aligned along the axis of the wind when viewed from several kilometers away (Figure 1.11(a)), the RCS are very irregular in close-up (Figure 1.11(b)).

This irregularity along the slopes is particularly marked in the *kaluts* in the Lut desert of Iran, carved out of soft clayey sediments. In the center of the desert, a zone that runs NNW-SSE for about 150 km, and measures 70 km, at its widest point (Figure 1.12), is made up only of long corridors that separate the clayey hills.

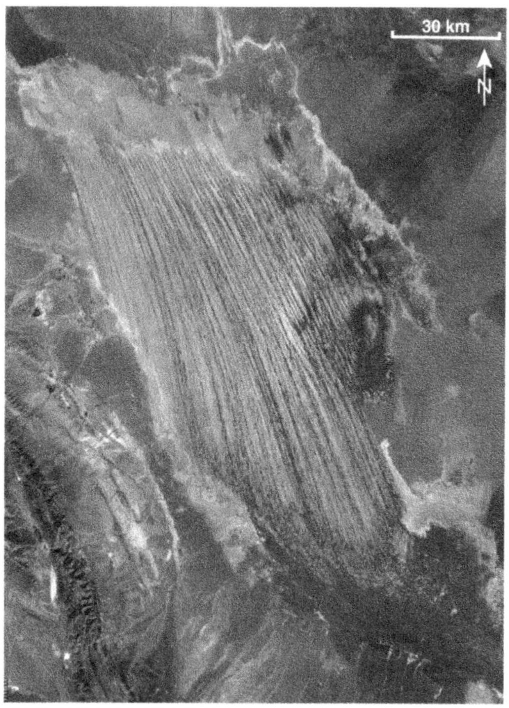


**Figure 1.12.** *The central region of* kaluts *in the Lut desert (Iran) (source: Google Earth). For a color version of this figure, see www.iste.co.uk/bourrie/deserts3.zip*

Satellite images clearly reveal the remarkable parallel arrangement of these structures, which can even be seen almost perfectly on medium-scale images (Figure 1.13(a)). On the other hand, close-up images (Figure 1.13(b)) show the magnitude of fluviatile dissection on the *kaluts*, which was a key contribution in the erosion of these structures.

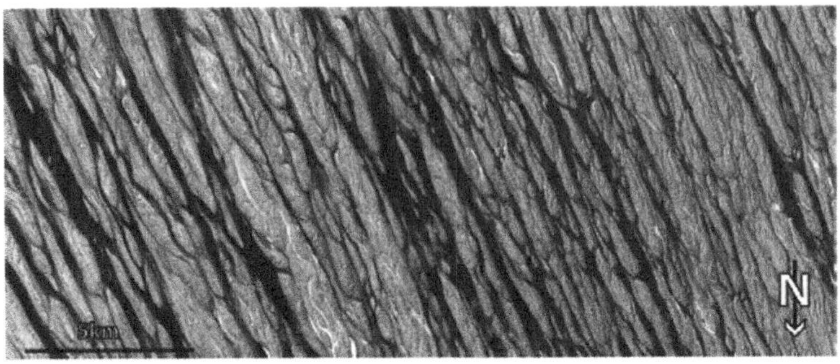


(a) In a medium-scale image, the structures remain parallel on the whole, even though they are not perfectly rectilinear.

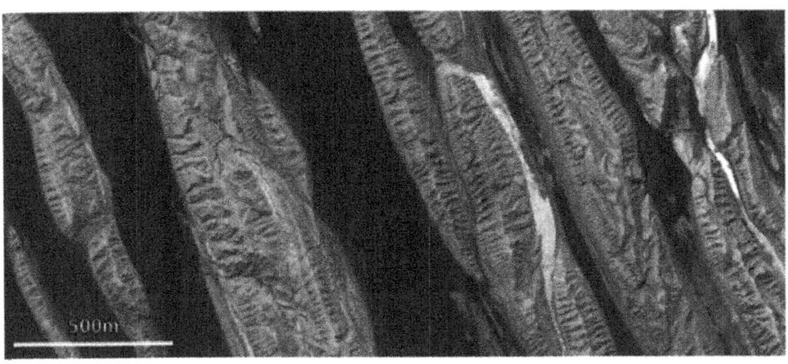


(b) In close-up images, the structures are completely broken up into badlands and traversed by hydrographic networks in directions that diverge from those of the *kaluts*.

**Figure 1.13.** *Ridge-Corridor Systems (RCS) in the Lut desert (Iran) (source: Google Earth)*

The number of processes, their alternation and their intensity make RCS into complex shapes that are largely beyond the scope of any work that studies aeolian desert formations alone; the time taken for them to be formed, which is hard to estimate, is tens or hundreds of thousands of years or even longer. Recent research (Sebe et al. 2011) has shown that these structures were also formed in periglacial environments in Central Europe, where very large forms (60 km in length, and 150 m high) have been studied in Hungary, in the western part of the Pannonian basin. It is estimated that these began forming about 1.5 million years ago.

### 1.2.3. *A mixed ablation shape: hydro-aeolian depressions*

The large RCS that were seen in the previous section are the first example of why aeolian and non-aeolian processes must combine for effective morphogenesis to occur. The creation of other hollowed-out formations, frequently seen in deserts, may result in different degrees of the same combination of closed depressions that cover large areas in many deserts around the world. These are topographically closed depressions, in the strictest sense, as the majority of deserts are endorheic and therefore have no flow toward the sea, and can be compared to hydrologically closed regions in a broader sense. These depressions vary in size and depth, from small depressions just a few meters across in diameter, spread out across the hamadas, to vast closed areas like the Kharga depression in Egypt, which extends across at least 12,000 $km^2$. There are several factors that contribute to the formation of these depressions, with tectonic movements playing a key role for the larger depressions. In other regions, they become lakes, like those in the East African Rift, or cease being closed depressions when a valley is carved out by the water body that runs through them.

In deserts, on the other hand, the absence of water makes endorheism possible and maintains this state. However, in all cases the natural tendency is for elements to be concentrated toward the center of the depressions, due to gravity and the run-off from the almost rare precipitation; this tendency can only be counteracted by an exporting agent that is free from the constraints of gravity, unless the filling up of the depression is counteracted by tectonic movement. Thus, the removal of material may occur in two ways: either through the dissolution of rock and the removal of material through the underground circulation of water, as in the case of karst depressions in soluble rock, or through an "upward" removal, which is only possible through aeolian transport. When closed depressions contain outcrops of soluble rock, it is not possible to conclusively identify their origins, which may be from a combination of processes. When the rock is not soluble, the "non-tectonic" depressions can only be aeolian depressions, i.e. caused by aeolian action. This is when the action of water plays an important role, as in the case of the RCS. Indeed, the alteration of the rock makes up for the low ablation capacity of the wind.

Many depressions of this kind, several kilometers wide, are found in the septentrional part of the Grand Erg Occidental. These are dug out in a very flat hamada, very slightly inclined toward the south (2 ‰ to 3 ‰), and indented by a few large wadis originating in the Saharan Atlas to the north, which disappear in the Grand Erg Occidental to the south. The Hamada is made up of continental, neogenic clay detrital deposits. They are not soluble in water but, on the whole, are not very resistant to erosion and are easily alterable; an important factor in the formation of a large number of depressions. The most remarkable closed depressions (Figure 1.14) can be found in the northwestern part of the erg, in the El Hobeur region.

## 1.3. Accumulation formations

When sand ceases to be displaced, it is deposited in different ways. *An aeolian accumulation of sand into a relief* is called a **dune**. It must have a certain slope as thick, flat sand layers cannot be considered to be dunes (see below). While this definition poses no problems for large structures, the same is not true for smaller forms. The minimum height of the relief has been arbitrarily fixed at 30 cm by (Glennie 1970; Cooke, Warren, and Goudie 1993) but only 10 cm by other authors (Wilson 1972). While sand is almost omnipresent in deserts, the accumulation of sand into formations like dunes is quite limited, given that dunes cover only about 20% of the Sahara. On the other hand, dunes are a constant feature of the popular imagination, where deserts are pictured as a series of undulating, endless strings of dunes. Geographic studies of representations of deserts show that dunes represent the majority of landscapes featured in writing on deserts, at the expense of regs, noted earlier (Roux 1996).

To remedy this overrepresentation, this book will not be limited only to dunes: going from the simplest to the most complex, we will study a series of non-dunal forms (unorganized), ripples, minor forms (of very small dimensions), sand accumulations that have no pronounced edges and sand accumulations with sharp edges. This classification has been chosen because dunes are often considered the only aeolian formation, even today, while active forms that see constant rearrangement of sand, or formations without sharp edges, are more often neglected even though they are found in large numbers. Further, this classification also has the advantage of presenting the structures in increasing dynamic order: all superficial aeolian formations are created through interactions between a fluid (air) and a solid (grains), making a loose aeolian formation whose shape can change. We will see further on that certain thresholds appear when the second factor (solid) modifies the behavior of the first (fluid) to a large extent.

(a) Very narrow bands of plateaus, the highest form arcs of a circle going from NNE-SSW in the north to ENE-WSW in the south east of the figure. Bands of intermediate plateaus follow a NW-SE direction. The Oued Namous (Grand Erg Occidental, Algeria) is completely superimposed on this aeolian structure.

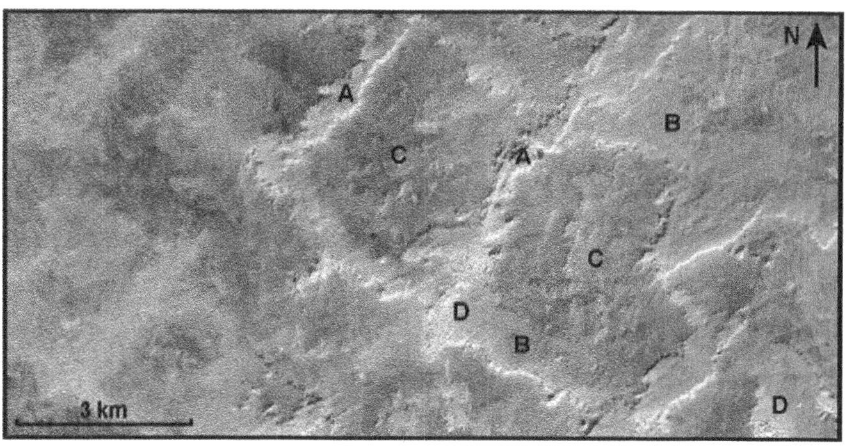


(b) The section framed in (a). A subhorizontal hamada is broken up to such an extent that only the NE-SW bands remain (A). A second phase of erosion has formed the NW-SE bands, orthogonal to the earlier bands, that are broader and less elevated (B). The ablation of the land between the bands has generated closed depressions (C) from where the material could only be exported by air. The dunes (D), very rare, are still along the bands made up by the hamadas.

**Figure 1.14.** *El Hobeur hydro-aeolian depression, to the west of the Oued Namous (source: Google Earth)*

### 1.3.1. *Amorphous accumulations*

The problem with the shift in scale between surface states and landforms is also found in the appearance of objects related to accumulation. The first of these, described here, are surface states that continue uninterrupted, in "dunal forms" for which the chosen criteria are essentially related to the dynamics of the deposits.

#### 1.3.1.1. *"Free" sand*

"Free" sand makes up the most discrete and also the most widespread superficial aeolian formation, while often posing a problem to infrastructure in dry regions. The denudation of the surface, due to the absence of continuous green cover, makes this sand omnipresent. However, although it is not very abundant, it appears to be so, masked by the irregularities of the soil. It can be moved if a sufficiently strong wind lifts it up. On the other hand, if the wind dies down or a large obstacle appears in the wind's path (especially one of anthropic origin), the sand can be deposited.

#### 1.3.1.2. *Sand sheet*

Sand sheets are fine (centimetric) sandy coverings that cover large areas. The sand in these sheets is often coarser, and sorted and re-sorted through deflation. As a result, these layers are important as they "protect" the surfaces. They are often likened to the "rust layer" that certain authors describe (Coudé-Gaussen 1991), whose color is linked to the iron oxides that cover the surface of the grains. The two terms are different however: the sand sheet designates the entire thickness of the deposit, with or without a rust layer; the rust layer, on the other hand, seen in the section on ablation forms, is the surface state alone, which may sometimes cover a sand sheet, but may form directly on a different surface (a clay, for example).

#### 1.3.1.3. *Nappages*

We define the term "nappages" here (a technique used in banking to form a glaze or layer) as continuous sandy accumulations of varying thickness, generally decimetric. They are conducive to plant growth as they store rainwater. However, conversely, they also form in areas where vegetation blocks the sand and, thus, it is often difficult to know which came first: the plants or the sand (Figure 1.15).

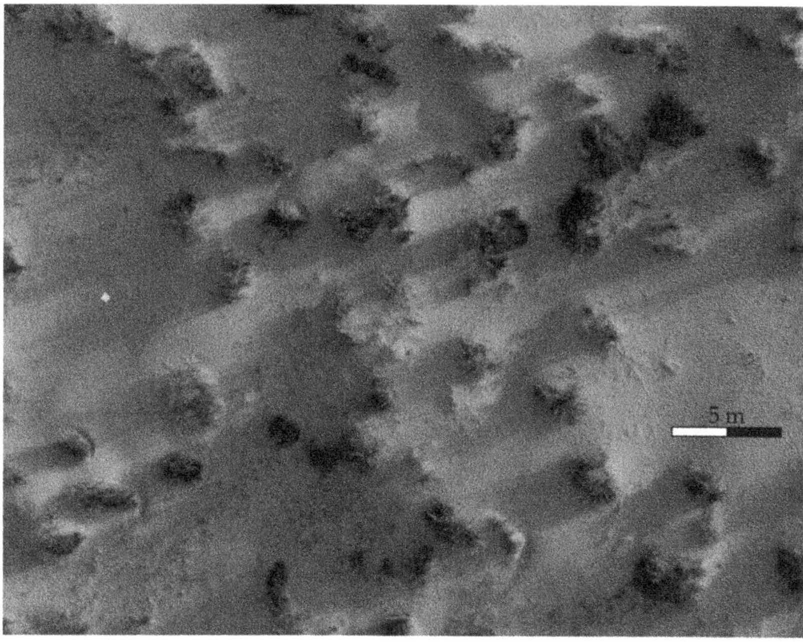


**Figure 1.15.** *Area with sand nappages in presaharan Tunisia. In a case like this, it is difficult to establish whether the formation of the sandy structure led to the appearance of vegetation or vice versa (image taken by Éric Delaître on a Pixy Paramotor). For a color version of this figure, see www.iste.co.uk/bourrie/deserts3.zip*

When climatic changes occur, all of these accumulations may be fixed by vegetation and subject to pedogenesis, leading to sandy formations that are more or less rearranged, whose aeolian origins can be demonstrated through various analyses (granulometry, heavy minerals, morphoscopy, exoscopy, etc.).

#### 1.3.1.4. *Sandbanks*

Sandbanks are sandy accumulations that may become very thick (several meters). They are classed under "amorphous accumulations" as they are formed on an obstacle or at its base: their formation and their size are, thus, more dependent on the obstacle and sand supply than on the wind. When formed on large relief features, they may be of considerable size; they may then form “clearer shapes” through the appearance of sharp edges, which turn them into dunes.

Their unique topographic position is often useful in understanding the direction taken by migrating sands (Figure 1.16): a thick sandbank at the base of a slope–which most often does not become as high as the slope–indicates that the accumulation takes place from the base upwards. The absence of any relationship between the sandbank and the peak or summit of the relief feature on which it is formed confirms this interpretation (sandbanks at the base of the most elevated points, for example). A thick sandbank on the top of the slope, which becomes thinner or disappears toward the base, indicates that accumulation occurs by the movement of sandy material below the relief’s peak. Each type of sandbank thus signifies opposing dynamics.

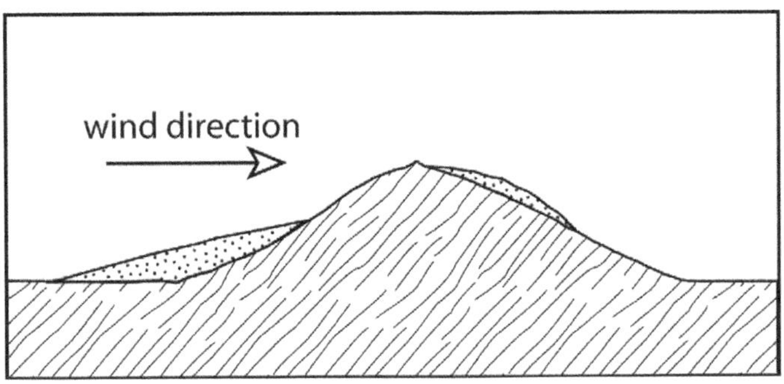


**Figure 1.16.** *Relationships between the topography, wind direction and aeolian sandbanks*

In mountainous or hilly regions, observations of the sandbanks cannot be generalized to form conclusions on the dominant direction of the wind on a regional scale. Relief can locally modify wind; it can completely shelter certain slopes from winds blowing in the opposite direction to that which the slope faces. On this kind of relief, the sandbanks are thus only valid indicators at a local level and no interpretation can be drawn as to a larger, regional dynamic. Certain infrastructural projects in arid regions have paid the price for this negligence or overgeneralization, as will be seen in the section on relationships between humans and the wind.

As soon as small relief features start appearing, sandbanks become a frequent formation. By definition they are stable forms, restricted by the topography, but can grow and stabilize depending on climatic variations and/or anthropic modifications to the environment, which may increase or diminish the supply of sand. It is difficult for the wind to completely erode them due to the sheltering effect the relief may provide. Their position against the relief, when these are prominent, often leads to modifications, or even varying levels of destruction by other processes linked to water or gravity (ravines, a pavement of protective blocks, etc.). Some of these structures may, therefore, be very old, with an evolution that is often very close to that of a glacis.

The "wind-sorting" effect discussed earlier plays an important role in all the superficial formations listed above. The wind tends to carry away fine, sandy material and concentrate the coarser, less mobile material until a more or less developed deflation layer appears. This layer protects the underlying formation from the wind's action and stabilizes it. When this layer is destroyed by anthropic action, it facilitates the movement of material again.

### 1.3.2. *Smaller aeolian formations*

#### 1.3.2.1. *Ripples*

There are several kinds of ripples, which have a more "visible" dynamic organization than the earlier formations.

*Wind marks* or *ripple marks* are found on the surface of the dunes (Figure 1.17). The granulometry, which can vary slightly across the ripple mark, is very similar to the granulometry of the rest of the dune.

They may also, however, be considered to be superficial changes to formations of another type (sandbank or dune). The ratio between the length of their waves and height varies between 12 and 50 – more generally around 17–18 – which makes it impossible to distinguish between these marks and ripple marks of aquatic origin. They are so short-lived that they cannot be used as indicators for the dominant wind direction, except through the use of multiple observations, as they only represent the cessation of the action of the last effective wind. On the other hand, they are typical of the surface of the most active aeolian forms.

**Figure 1.17.** *Ripple marks on the two faces of a sif (north of the Grand Erg Occidental, Algeria). For a color version of this figure, see www.iste.co.uk/bourrie/deserts3.zip*

A relationship is seen between the sand granulometry and the size of the ripples (Callot 1987): the coarser the sand, the larger the ripples can be. In Issaouane n'Tiffernine (to the east of the Algerian Sahara), these ripples can be 30 cm high, with wavelengths between 2 m to 5 m large, extending over a length of 20 m or more (Figure 1.18(a)). The largest of these can be seen in high-definition satellite images, as in the Western Desert of Egypt, to the west of Luxor. Here, in the Kharga depression the ripples can be as long as 60 m (Figure 1.18(b)).

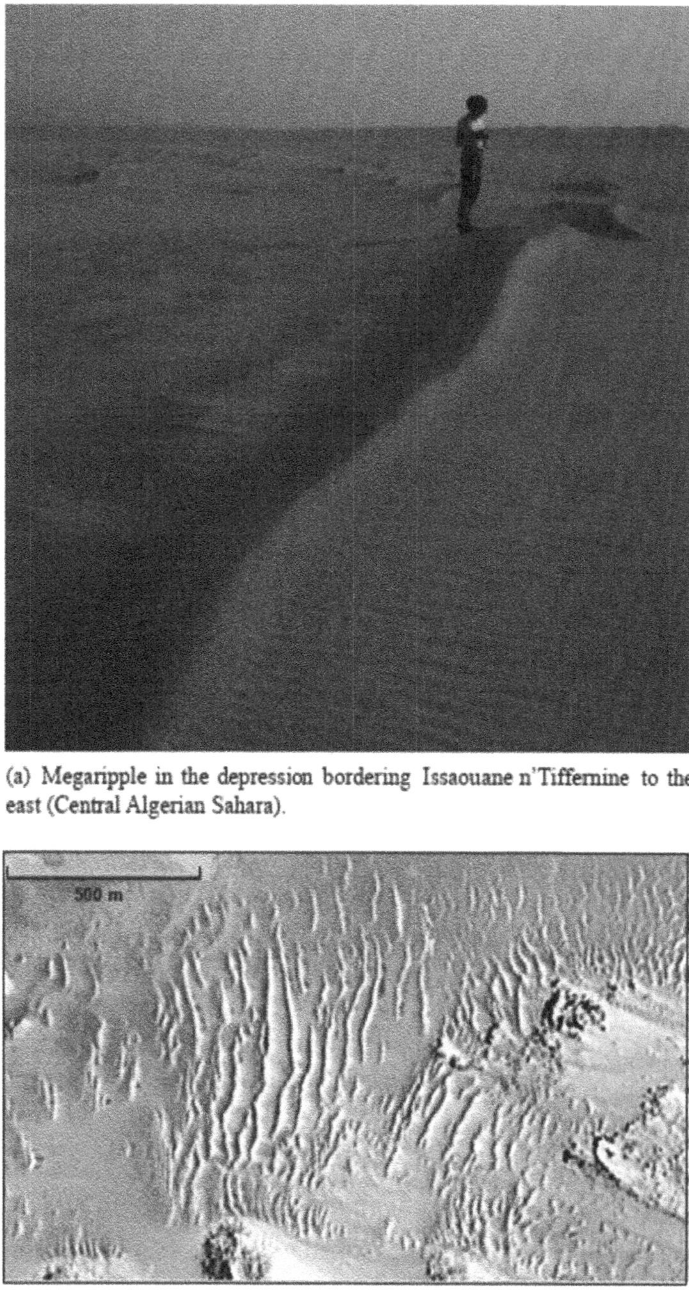

(a) Megaripple in the depression bordering Issaouane n'Tiffernine to the east (Central Algerian Sahara).



(b) A field of megaripples in the Kharga depression (Western Desert, Egypt). Taken from a Google Earth GeoEye image.

**Figure 1.18.** *Megaripples. For a color version of this figure, see www.iste.co.uk/bourrie/deserts3.zip*

The majority of these large ripples have no sharp edges as they are formed of very coarse material (up to 2 mm), which does not allow the formation of a pointed crest. Given their granulometry, their formation has been attributed to the more violent paleo-winds (perhaps toward the end of the Würm interglacial). This would make them legacy forms, or forms that formed over a very long period (Bagnold 1960). But we have also seen that these ripples re-form when wiped out by tire-tracks. This would indicate that they are still functional, can form anew quite quickly and that their position is stable (at least for the cases observed). In the Grand Erg Occidental (Algeria), we have seen identical ripple marks in coherent, soft Holocene lacustrine sediments. It is possible, therefore, that these ripples can be formed by the further ablation of furrows formed in a deposit, rather than loose material being brought in and deposited.

These large ripples are typically the largest possible form, both in terms of the process through which they are formed, as well as their shape. It is possible to ask the following question: Are these formations dunes, since some in the Kharga depression exceed 30 cm and since some authors consider that the term "dune" can be applied for structures that exceed 10 cm in height? This difficulty may be resolved by looking at the granulometry (Cooke, Warren, and Goudie 1993): in structures of this dimension, those with a sand granulometry of 120 μm to 250 μm can be classified as dune forms, while those with a coarser granulometry (>500 μm) are classified as ripples.

#### 1.3.2.2. *Aeolian formations on obstacles: nabkhas and rebdous*

These sandy accumulations involve the presence of a pre-existing obstacle that blocks the migration of the sand. Thus, by definition they are fixed in position, though this does not exclude the possibility that they may be modified through changes in shape and/or direction. These could be very useful in understanding the dynamics of the landscape, especially in regions on the borders of the desert.

**A nabkha** (Figure 1.19) is the best-known and most frequently occurring of these formations. It is a small, sandy accumulation that stretches out in the direction of the last effective wind, formed around and on the leeward side of an obstacle. This obstacle may be a clump of plants in semi-arid or arid regions, or a block of rock in hyper-arid regions; or even an engine or gearbox in inhabited regions. Above all, when it is small, a nabkha may change direction very quickly (in a few minutes) and, consequently, it is not indicative of the annual resultant migration of sand.

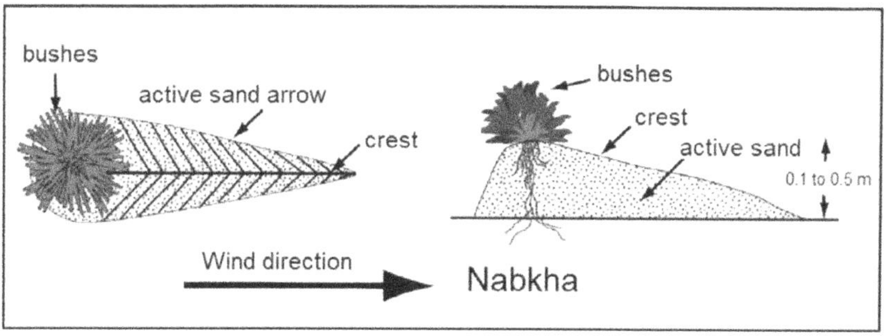


**Figure □.□□.** *Diagram and cross-section of a nabkha*

In certain sections where there is a large supply of sand and vegetation is dense enough, there are fields of nabkhas that almost form a chain, which can increase in thickness when there is a conducive combination of circumstances: accumulation promotes water being trapped in the sand, which is conducive to plant growth and leads to further accumulation. In this case, the nabkha then becomes a formation made up of sandy layers, as can be seen in Figure 1.15. **A rebdou** (Figures 1.20 and 1.21) is formed around larger plants than those that result in a nabkha, like a shrub, often a tamarisk. The sand is trapped in the branches and the roots, smothering the plant, which then grows taller to emerge from the sand. This sand, which hardens around the roots, becomes coherent and the rebdou acquires a slope that is steeper than the slope formed by active sands. It may be 10 m high. When the plant dies, the structure is attacked by corrasion and can sometimes produce a streamlined form like a yardang (see above).

Sandbanks and accumulations on obstacles, especially nabkhas, reveal one of the biggest constraints of building installations in arid regions: **the installation of any obstacle (buildings, road embankments, etc.) modifies the dynamics and blocks sand, often leading to the structure being covered by sand.**

All of these **small aeolian formations**, especially those that form around plants, the most frequently occurring, are called aeolian forms around obstacles (A□□) and they may appear to be of little significance, at least with respect to their size. However, certain features make them of greater interest than others: they cover vast areas, especially in semi-arid regions or subarid regions where they may be the main sandy formations. Recent research (Ouerchefani 2012; Venard 2012) has shown how useful they can be: they are remarkable indicators of subtle changes in the environment, especially the advance (or retreat) of desertification. Nabkhas, for example, which are initially active and changing, can stabilize and be covered with a more resistant layer due to biological activity; sometimes, if the initial plant around which it formed dies, the nabkha may be eroded by wind or other agents – such as herds of animals – and may disappear. They thus have a short lifespan, most often spanning a year or perhaps a decade, which is governed by the lifecycle of the plants, climatic changes and/or anthropic pressure.

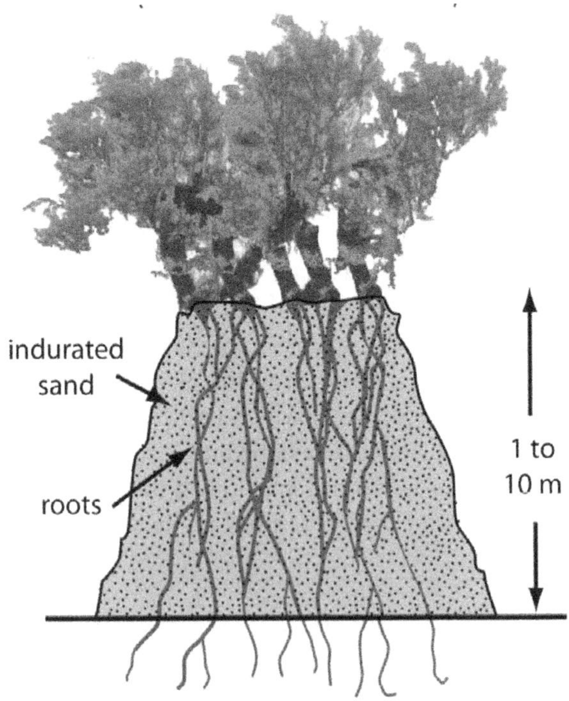


**Figure 1.20.** *Schematic cross-section of a rebdou*

**Figure 1.21.** *Rebdous at the foot of Issaouanne n'Tiffernine (Algeria). The structure to the right, about 10 m high, is still stable; the structure on the left is being demolished by the wind following the death of the tree around which it formed. For a color version of this figure, see www.iste.co.uk/bourrie/deserts3.zip*

### 1.3.3. *Sandy accumulations without sharp crests*

A large portion of the sandy area is not made up of dunes with sharp, pointed crests, but of less "obvious" shapes, lacking angularity. The absence of these sharp crests in a sandy accumulation may have several causes that could occur together:

– the very small size of the structure, too small to block the wind and form a sharp ridge, as is the case for the small aeolian formations seen earlier;

– granulometry that is too coarse. Coarse sand, which is less mobile, rarely forms sharp edges, as we saw in the case of the giant ridges;

– the development of green cover, which slows the aeolian dynamics and prevents the formation of a crest.

Dune formations without sharp crests are more difficult to study than dunes with pronounced crests, as they do not have a clearly defined form by definition. For smaller structures, we can use the term "heaped dune" to designate structures that are not large enough to disturb air flow sufficiently to create a "dynamic" organization of the dune. These can be seen in aerial photographs of regions with mono-directional winds, mixed with larger barchan-like dunes, with all the intermediate forms showing the change from one form to another (see Figure 1.26).

In most cases, if the dunes are large and if the region is too arid for vegetation to intervene, granulometry plays an essential role in their formation. These dunes cover vast areas, like the sand seas in the eastern Sahara, to use a local term. It is often impossible to know whether the formations are simple or a polygenic form with a more complex "history". Furthermore, when the slope is very small, it is sometimes difficult to differentiate these dunes from sandbanks.

**Zibars** are an excellent example of the difficulty in studying these formations. These are dunes that are very frequent, for instance, in the Erg of Fachi-Bilma (Chad-Niger, Figure 1.22) where they make up the substrate of the active formations with finer granulometry (Mainguet and Callot 1978). With a low height ($\leq 5$ m), small slope (their wavelength ranges from 100 m to 300 m), and sometimes asymmetric, these formations have a marked bimodal granulometry, a mix of coarse sand and very fine sand. This is attributed to the wind sorting a material that was originally very heterogeneous (Warren 1972; Cooke, Warren, and Goudie 1993): the wind exported the medium sand but could not greatly displace the coarser sand that protected the fine material.

### 1.3.4. *Dunes with sharp crests*

#### 1.3.4.1. *Sif – the basic component of dunes with sharp crests*

Dunes with sharp crests are the best-known structures as they are the most spectacular. They result from a stronger interaction between wind and sand. If the latter is too abundant and/or if the aeolian formation becomes too large, clearly defined crests start appearing. The terminology used for dunes is still variable and the definitions given here could differ from author to author. If we consider that the clearly defined crest is a unique feature of these dunes, they may be considered to be a more or less complex combination of crests. This is why the terminology used here is constructed based on the "crest" taken in isolation, called a [illegible] ([illegible] in Arabic), which can be defined as *[illegible]*. Sifs are classified according to their curvature and their combinations, since they are rarely isolated, except in their final limiting form, the barchan.

(a) Google Earth images of zibars in the Erg of Fachi-Bilma, 60 km ENE of Fachi. These highly sinuous undulations have such small slopes that there is no clear shadow, like those seen on the silks visible at the top of the illustration.

(b) View of the soil on the zibars in the same region. The ripples in the foreground are typical of coarse sands. The caravan makes it possible to perceive the undulation that is not very marked. In the distance, the edge of the silk.

**Figure [illegible].** *Z ibars in the Erg of Fachi-Bilma (Chad-Niger). For a color version of this figure, see www.iste.co.uk/ bourrie/ deserts3.zip*

Contrary to what may be assumed, most of these forms with clearly defined crests are fixed dunes. In active dunes, the way that aeolian grains are placed – by simply falling, by rising on the windward faces, collapsing on the windward faces – when combined with ablation phases, form a variety of sedimentary structures based on the type of dune.

However, a few common characteristics can be identified despite the variations:

– the sand bed is often inclined, especially on the leeward side (the slipface) whose angle is greater than 30°;

– the magnitude and frequency of its discontinuities (truncations), which vary in scale, separating chains of beds. These often have a low angle of incline and correspond to the alternations between the accumulation phases and ablation phases, which may be linked, for example, to the changes in wind direction on a silk, or to the superimposition of structures in regions with mobile dunes. These discontinuities may be complex, even in structures that appear simple, such as sifs (Figure 1.23).

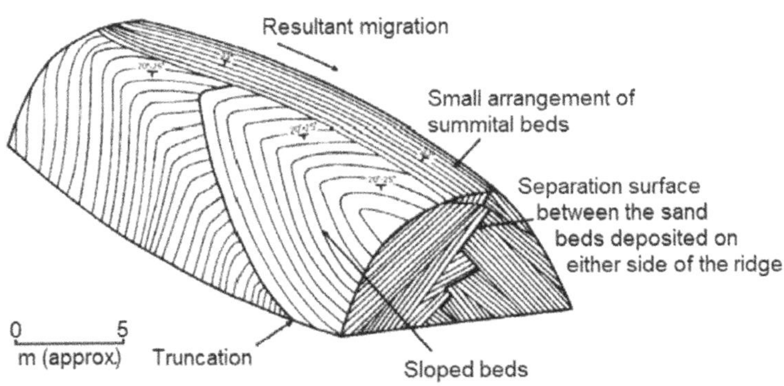


**Figure 1.23.** *Stratification within a sif (according to Tsoar (1982)). Note the number and size of the truncations. The resultant of the migration is only true in most cases at the level of the free sand grains, not at the level of the dune*

1.3.□.2. *Barchans*

A barchan (Figure 1.2□) is a highly mobile dune and the exception that proves the rule. It is a [illegible] *curvature of the sif under the influence of a monodirectional wind.* The definition of a barchan emphasizes two characteristics that are strongly interlinked. In regions with strong prevailing winds, the sifs curve inwards until they form a crescent shape. The windward side of the barchan has a gentle slope. The slipface has a "crumbling" slope, due to the collapse of the sand (in small streams) that accumulates at the top of the slope that is sheltered from the wind (Figure 1.25).

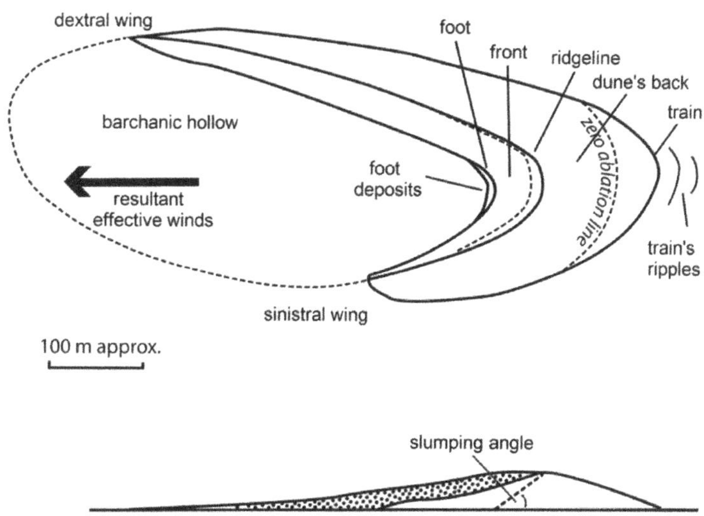


**Figure 1.24.** *The chief components of the barchan (plane and cross-section, according to Oulerhi (1992)). The structure shown here has a dextral wing longer than the sinistral wing*

A barchan is the *only truly mobile dune*, moving in a direction that corresponds to the bisectrix of the angle formed by its two wings (Figure 1.26), which are called the dextral, the wing seen to the right of an observer on the crest of the dune when their back is to the prevailing wind, and the sinistral, the wing seen to the left.

Barchans only exist in regions where the wind has a dominant direction (with the exception of the tiny structures that form during a single sandstorm, which disappear afterward). They have long been considered a vehicle for sand, with the displacement of the structure being seen as its function.

However, this structure leads to the movement of sand slowing down, which would migrate much faster if it were not incorporated into the barchan, where it remains stationary for a duration, which becomes longer when the dune is large. The barchan is thus a **mobile stock of sand** in an environment where sand migrates rapidly, but is too large to be able to migrate as quickly in its entirety.

**Figure 1.25.** *Crumbling slope of a barchan, near Tarfaya (Morocco). The gently sloping part of the dune, facing the wind, is fluid because of blowing sand, whose height is not more than a few centimeters. The sand arriving on the slipface accumulates on the steep slop until the slope is so great that a stream of sand is produced. To the right, the stream of sand is in action. To the left can be seen an earlier stream, in the middle of which are the traces of an even older stream. For a color version of this figure, see www.iste.co.uk/bourrie/deserts3.zip*

Many researchers and those with a passion for deserts are interested in barchans because of their apparent simplicity of form and function. They are formed from circular or ovoid heaps of sand with no clear edges, sometimes called shield dunes, which are formed (as we saw) when the wind cannot transport all the available sand. All the intermediate forms exists between this "shield", with no defined shape, which belongs to the group of forms with no sharp edges, and a proper barchan (Figures 1.25–1.27): the shape is formed through the enlarging of a relief form that promotes the deposition of sand on the slipface; the shape then becomes asymmetric, with the incline of the slipface slope increasing until it starts crumbling and wings start forming.

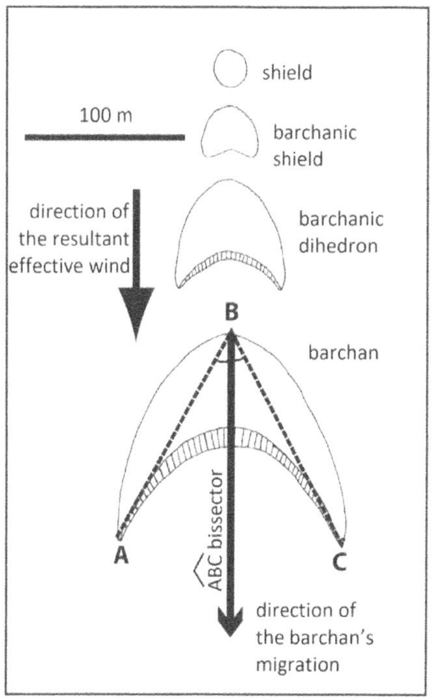


**Figure □.□□.** *The transition from a shield, a shape with no defined edges, to a barchan. The dimension of the structures, which varies depending on local conditions, ranges from a few tens of meters to hundreds of meters*

When barchans reach a certain size (at least 100 m in width), they are of great interest in terms of geomorphology and planning: in the absence of any meteorological data on the winds, they are an indication of the results of the wind and, therefore, of the direction of the sand migration, which corresponds to the direction in which they are displaced.

### 1.3.□.3. *The displacement of barchans*

The higher the barchan, the more slowly it is displaced, because, as a barchan grows, the wind acts on an area that grows as a squared function, while the volume increases as a cubed function. The ratio between the surface exposed to the wind and the volume to be transported reduces, thus slowing displacement. This feature was earlier used by certain inhabitants of the Atlantic Sahara, who sought to protect their goods from pillagers or tax collectors and would hide them in front of a barchan and collect them after the threat had passed (Oulerhi 1992)!

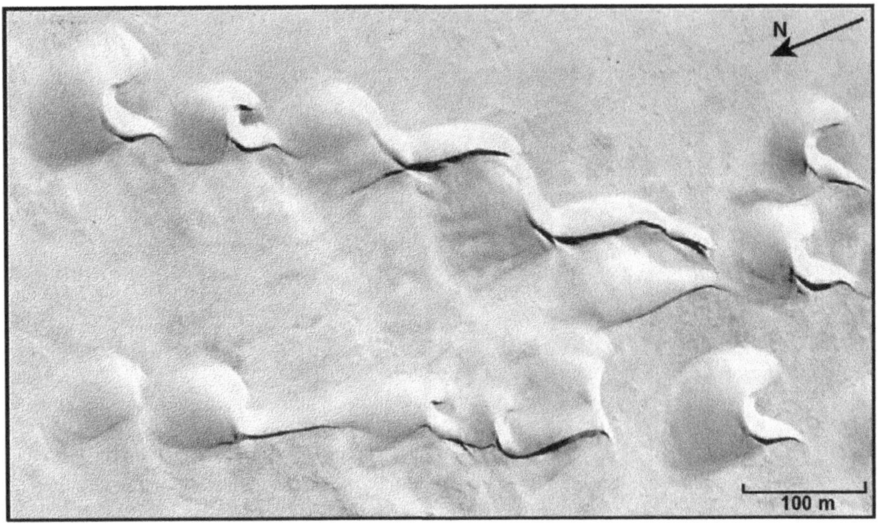


**Figure 1.27.** *Barchan formations to the east of (NW Mauritania) with all the intermediate shapes between the shields (bottom left) and the barchans proper (source: Google Earth)*

In NW Mauritania (Figure 1.28), Coursin (1964) demonstrated that the rate of displacement of the barchans follows a simple formula:

$$V = 300/H, \quad [1.1]$$

where $V$ is the average annual displacement of the barchan in m/yr and $H$ is the height of the barchan in meters.

This would mean that a dune with a height of 30 m would move by $300/30 = 10$ m/yr, while a dune with a height of 10 m would move $300/10 = 30$ m/yr; thus, the smaller dunes advance faster than the larger ones.

**Figure 1.28.** *Barchan, called "Dune Coursin", close to Boulanouar (NW Mauritania). Tracked since the 1950s, this has provided information over a half-century on the displacement of barchans*

This formula cannot be generalized, as the rate varies from one region to another depending on the wind regime. Other formulae have been proposed by many authors for other parts of the world (Oulerhi 1992). Based on the texts compiled by Oulerhi, it is possible to propose a transport coefficient that is based on the displacement rate of the barchans (Table 1.1). It is computed by taking the product of the height of the barchans and their displacement rate, and thus corresponds to a cross-product from Coursin's formula. If $V = 300/H$, then the coefficient $H * V = 300$. This table shows that there are significant differences between coefficients, depending on the region, which can be explained by a difference between winds. However, there are also differences for the same place between authors, with figures that may be aberrant. The use of high-definition satellite data today makes it possible to calculate the displacement with much greater accuracy. For instance, Venard (2012) has proposed, for the Boulanouar region, a coefficient of 290, computed through processing an image of 30 barchans. The results presented here are therefore likely to be greatly refined in coming years.

Barchans may be asymmetric, with one arm longer than the other. They are called "sinistral" if the longer wing is to the left of the bisetrix of the two wings, observed from the peak of the dune, and "dextral" if it is to the right. This elongation used to be interpreted as the action of a secondary wind. However, Clos-Arceduc 1965 invalidated this hypothesis, showing that barchans situated very close to one another had elongated arms in opposite directions. This asymmetry can be explained by the rate of displacement, as discussed above: a smaller (and therefore faster) barchan catches up with a larger one and "melts" into one of its arms (Figure 1.29). This then forms a second-degree complex form, while a single barchan is a first-degree dune, i.e. it is not formed by the combination of other elementary structures.

| Region of migration (/m) | Location coefficient (/m yr$^{-1}$) | Height (/m$^2$ yr$^{-1}$) | Rate | Transfer | Ref. |
|---|---|---|---|---|---|
| Peru | Pampa de la Joya | 1 to 7 | 32 to 9 | 32 to 63 | 1 |
| Peru | Pampa de la Joya | 2.1 to 3.9 | 28 to 16 | 59 to 62 | 2 |
| California | Algadonos | 0.5 | 5 | 2.5 | 3 |
| California | Algadonos | 6 | 20 | 120 | 4 |
| California | Salton Sand Sea | 3.1 to 8.2 | 27 to 14 | 83.7 to 114.8 | 5 |
| California | Tulewash, Salton | 11 | 13.4 | 147.4 | 3 |
| California | Guerrero Negro | 7 | 18 | 126 | 6 |
| Egypt | Abou Moharic | 6 to 14 | 8 to 5 | 48 to 70 | 7 |
| Sinaï | El Arich | 1.9 | 13.1 | 24.9 | 8 |
| Sinaï | El Arich | 3 | 6.4 | 19.2 | 8 |
| Sinaï | El Arich | 4 | 6.2 | 24.8 | 8 |
| Saudi Arabia | Al Hasa | 20 | 15 | 300 | 9 |
| Saudi Arabia | Erg Jafourah | <30 | 15 | <450 | 10 |
| Mauritania | SE of Cap Blanc | 13 to 17 | 63 to 18 | 819 to 306 | 11 |
| Mauritania | Boulanaouar | 10 | 25 to 30 | 250 to 300 | 12 |
| Mauritania | Boulanaouar | 10 | 30 | 300 | 13 |
| Mauritania | Boulanaouar* | | | 290 | 14 |
| Atlantic Sahara | Laayoune Province | 8 to 9 | 36 to 32 | 288 | 15 |

* Thirty dunes; 1. Finkel 1959; 2. Hashtenrath 1987; 3. Norris 1966; 4. Smith 1970; 5. Long and Sharp 1964; 6. Inman 1952; 7. Beadnell 1910; 8. Tsoar 1974; 9. Abolkhair 1981; 10. Fryberger et al. 1984; 11. Sarnthein and Walger 1974; 12. Capot-Rey 1963; 13. Coursin 1964; 14. Venard 2012; 15. Oulerhi 1992.
The transfer coefficient is the product of the height $H$ and the migration rate $V$ (equation [1.1]).

**Table 1.1.** *Computation of the transport coefficients based on the original table proposed by Oulerhi (1992), using bibliographical data from Thomas (1992), with additions (marked by asterisks) and modified*

### 1.3.5. *Combinations of sifs*

These structures are called second-degree structures as they are made up of a combination of sifs arranged in a linear or roughly radial manner.

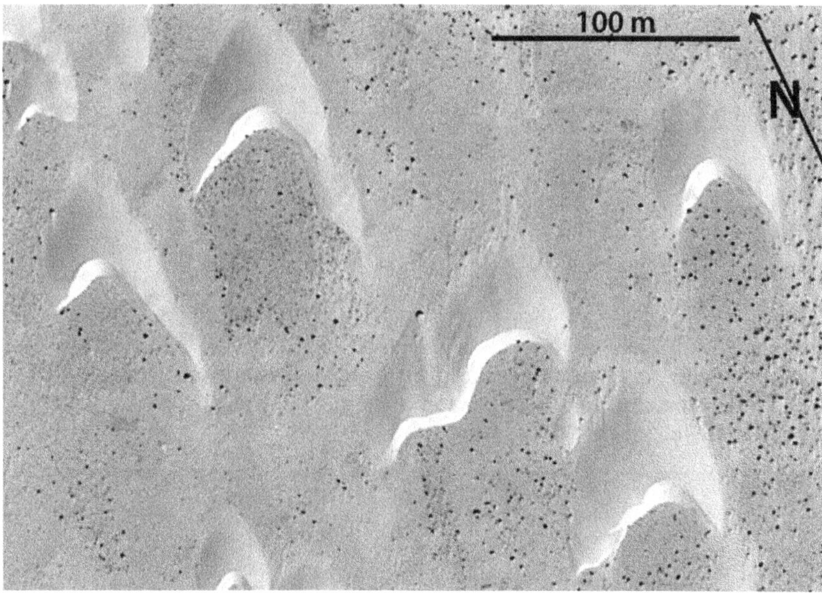


**Figure 1.29.** *Barchans with wings in opposite directions 8 km south of Tarfaya (Moroccan Sahara) (source: Google Earth). North is to the top. Several stages of integration of the smaller barchan into the larger can be seen: on the central barchan the two structures can still be differentiated in a sinuous dextrous arm; on the other hand, the barchan further west has an elongated, rectilinear sinister arm where all traces of the dune that formed the arm have disappeared. On the top left, there is a shield dune*

#### 1.3.5.1. *Silks*

A silk is a *rectilinear, sandy cordon, which is flat overall and more or less sinuous in close-up, due to the end-to-end alignment of sifs* (Figure 1.30).

Some authors, especially English-speaking authors, do not take this alignment into account and use the term *sif* for a silk, without using any term to designate the basic element (which we are here calling a "sif"). The term "longitudinal dunes" is sometimes used for a sif, but this must be used with caution: it implies that the structures are aligned along the resultant of the effective winds, which is often not the case. Some English-speaking authors also use the term *linear dunes* (Cooke, Warren, and Goudie 1993) for structures similar to silks. This term has the advantage of not involving any hypothesis on the genesis of the structures, especially their longitudinal character, however, it is used for structures of widely varying sizes.

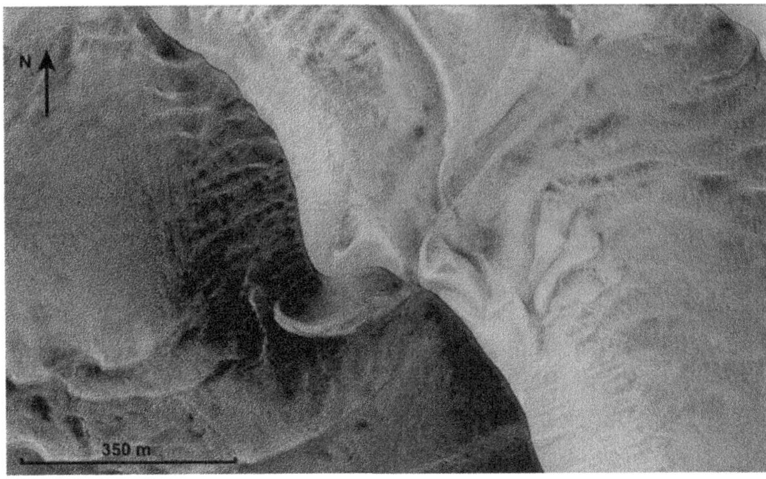


**Figure 1.32.** *Ghourd with four edges converging on the summit, or top, in the Namib desert (source: Google Earth). For a color version of this figure, see www.iste.co.uk/bourrie/deserts3.zip*

Observations from various deserts show that **ghourds are dunes whose positions are remarkably fixed**. This does not imply that they are totally immobile, however, while their crest may be displaced a few meters, depending on the wind, their base is not affected at all. This stability can be seen on a historical scale. For example, in northwestern Algerian Sahara, in the Grand Erg Occidental, the foot of a ghourd has been about 10 m from the village of Taghit for centuries (Figure 1.34).

**Ghourds, which are the largest dunes in the world by volume, show that the image of mobile dunes is most often erroneous.** It is only the barchans and other small structures that are displaced in regions where winds follow a markedly dominant direction, causing a significant migration of sand, which, on a larger scale, allows the displacement of the dunes. In other regions, dunes may be formed and displaced slowly in areas that are otherwise stable, such that one can navigate through certain ergs, using aerial photographs dating back several decades. This stability implies that it is possible for aeolian formations in the same region to be affected by a variety of paleoclimatic changes and these large dunes could, in the future, be a promising source of information, if researchers are able to extract soil from the depths of the fixed parts and date them using Optical Stimulated Luminescence dating (OSL) techniques.

(a) Oblique aerial view.

(b) Ground view.

**Figure 1.33.** *Large, complex ghourds at the meridional end of the Issaouane n'Tiffernine (Algerian Sahara), surrounding a Holocene paleolake. In the view from the ground (b), the dune on the left is about a hundred meters high and that on the right is about 300 m high. The contrast between silks and ghourds is very visible in both photographs. For a color version of this figure, see www.iste.co.uk/bourrie/deserts3.zip*

**Figure 1.34.** *The ksar of Taghit, at the foot of the ghourds on the western border of the Grand Erg Occidental. For a color version of this figure, see www.iste.co.uk/bourrie/deserts3.zip*

#### 1.3.5.3. *Aklé*

An aklé (or akle) is a *planar combination of sifs, often with a pronounced curvature, arranged in groups that are more or less rectilinear and joined-up, appearing as a "lacy" formation in aerial photographs* (Figure 1.35).

Some authors (Cooke, Warren, and Goudie 1993) consider that aklés are chiefly transversal forms, i.e. forms that are roughly perpendicular to the resultant of the effective winds. However, we must be careful: many aklés are not arranged in this way, or are organized in two, almost-orthogonal directions, (Figure 1.35(a)) or have no real dominant direction (Figure 1.35(b)). When the aklés are narrow, they are often very difficult to traverse. They, are often neglected, a little like the regs, as they do not have a pronounced "typical" form and are not very elevated. However, they cover vast areas, sometimes superimposed onto other sand forms in a complex formation (see below).

(a) Aklé in two dominant directions of the sifs in the Grand Erg Occidental.

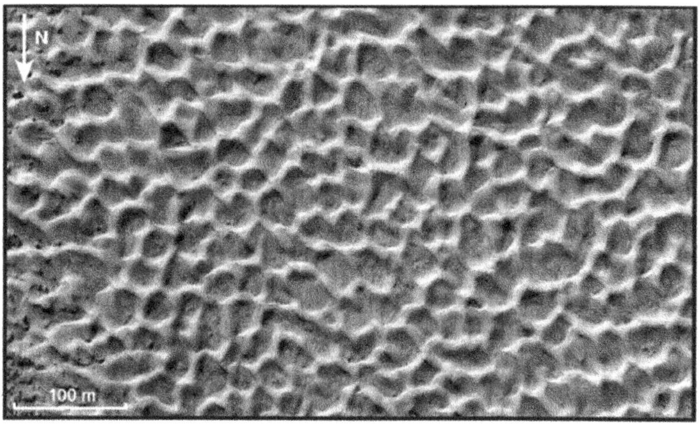


(b) Aklé south of Brezina (northwestern Algerian Sahara) with sinuous crests that are not very sharp but are very narrow.

**Figure 1.35.** *Two types of Aklés in Algeria (source: Google Earth)*

### 1.3.6. *Non-sandy aeolian accumulations*

While aeolian accumulation formations are most often made up of sand, where silica is the chief material in the precise granulometric classes described above, there are also some other accumulations. These differences could be for several reasons, see sections below.

#### 1.3.6.1. *Accumulations of "sandy", non-siliceous material*

In granulometric classes, comparable to those of active, siliceous dunes, these formations may be composed of another mineral. This is very rare in arid regions, apart from accumulations of "gypseous sands" situated on the leeward side or slipface of large, salty depressions that yield large quantities of evaporites. The material, which is seemingly similar to siliceous soil is "lighter": when examined under a binocular microscope or under a scanning electron microscopy (SEM), it is seen to have a flake-like structure of gypsum agglomerates separated by many voids, which explains the low density of the rock.

Apart from a few, rare sites, these accumulations are quite limited. For example, they exist in Algeria, in the south Oran high plains around the Chott Chergui, however, they are no higher than a few meters. The best-known site is the White Sands National Park in New Mexico: as the gypsum here is very pure, the sands are perfectly white and form dunes that are quite high, almost appearing to be made of snow.

Similarly, barchan formations that have the highest rates of displacement in the world have been studied in the Bodélé region in Chad (Vermeesch and Leprince 2012). This speed is related to the exceptional nature of the "sand": it is almost exclusively composed of diatomite whose apparent density is very low – 0.2g/cm$^3$ to 0.4g/cm$^3$ when it is pure (Inglethorpe 1993) – because of the number of voids in the rock. The material transported is, thus, so light that it is transported much more rapidly by the wind, with displacement coefficients that are close to 600.

On the other hand, accumulations that are much more widespread are formed, as a result of the granulometry of the accumulated material, which is much finer than siliceous sands.

#### 1.3.6.2. *Accumulations of fine material located very close or close by: clayey lunettes and gypseous sandbanks*

These "aeolian" formations have the distinctive feature of being totally stable once constructed, due to the manner in which they are formed. They form in proximity to a sabkha, a salt lake that alternates between flood phases and dry phases. During the periods when it is inundated, the clays at the bottom of the sabkha absorb salt water. When they are dry (a result of evaporation), micro-crystals of salt form between the sheets of clay. This micro-crystal process separates the sheets, deflocculating the clay and transforming it into an extremely light powder, sometimes called "pseudo-sand".

**Lunettes** are formed because these pseudo-sands, which are easily transported by the wind given their low density, are displaced in ways that change very quickly: they either hit the sand and quickly break up into very fine particles that are carried away in a suspension (see above) over very long distances, or they encounter plant cover on the edge of the sabkha that traps them because of the roughness. Thin banks of "sand" then form. When the first rains fall, the water dissolves the salt crystals and carries them away as a solution. The clay flocculates, its sheets "stick" together again, and it once again becomes coherent, forming a compact accumulation that the wind can no longer move. Over a long period, a lunette is formed: a hill with a very gentle slope on the leeward side of the sabkha, stretching parallel to its shore. One of the most remarkable examples of a lunette is found by the Sebkha d'Oran, in Algeria, stretching over some 30 km, with a width of 500 m to 1500 m; it is over 60 m tall in its eastern part, diminishing toward the west as the supply of material reduces (see Figure 1.36).

**Gypseous sandbanks** often have the same origin as lunettes: a sabkha where gypsum is present in a dissolved form. However, they may also be formed from gypseous outcrops in sedimentary series or superficial formations, or in deposits on the beds of wadis with selenitic waters. Weathering breaks up the gypsum into many thin sheets, with a diameter measuring a few millimeters or smaller, which are very easily moved by the wind given their mass–surface ratio, which is very conducive to aeolian transport (see section I.2.1, p. xx). The gypsum "flakes", transported as a discontinuous suspension, will get deposited on the leeward side of the sabkhas. The precipitation will tend to make them coherent again, so much so that they form hard and infertile gypseous crusts. The aeolian origin of these formations was demonstrated by Coque in 1962.

The study of the types of accumulation discussed above shows that they all have one common feature: they are tributaries of a large and nearby supply of material that is moved by water in one of the steps that leads to their formation. This explains why these formations are absent or underdeveloped in truly arid regions where water is extremely rare. Thus, they largely exist in semi-arid regions, as seen in the two examples discussed here. On one hand, they may influence human activities: sandy lunettes often have very rich, non-saline soil without coarse elements, where agriculture may be practiced. On the other hand, gypseous sandbanks are very unfavorable for agriculture.

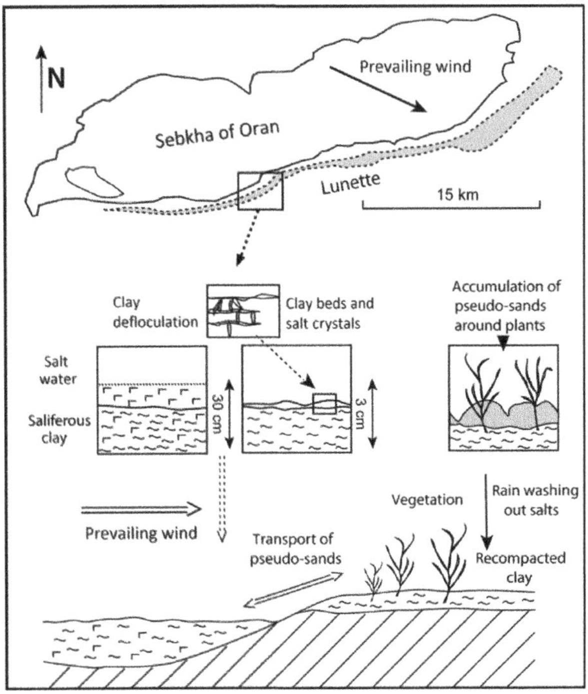


**Figure 1.36.** *The lunette at the Sebkha d'Oran (sketch) constructed leeward of the prevailing winds in the western sector*

### 1.3.6.3. *Accumulations of distant fine material: desert loess*

When aeolian material is very fine, it may be transported over large distances (see above), with the length traveled depending on the granulometry. The broken-up pseudo-sands in sabkhas, as well as many other materials, called desert aerosols, may travel great distances in a suspension, covering as much as hundreds or thousands of kilometers. They then get re-deposited in favorable areas, characterized by significant roughness, chiefly due to extensive grassy or bushy green cover. This plant cover forms an obstacle for the material by preventing the wind from blowing along the ground (see section I.1.1). These processes of capture form deposits with a high silt-clay proportion. The manner in which these deposits form is quite similar to the formation of periglacial loess (Figure I.I.3, p. xviii) and, consequently, these are called "desert loess" even though their granulometry is a little coarser than that of the periglacial loess.

The necessary presence of green cover results in these deposits being much more frequent along the borders of arid regions than within true deserts. However, the presence of water, especially artesian flows, may allow vegetation to appear in the desert itself, regardless of precipitation. A striking example of this is the meridional part of the Kharga depression in Egypt, in which the upsurge of artesian water has allowed plant growth and agriculture for millennia. In a hyper-arid context, the rapid accumulation of fine and homogeneous material over several meters may be interpreted (Crépy 2016) as loess, deposited through this green cover and trapping desert aerosols, subsequently eroded into yardangs when the water flow ceases (Figure 1.10).

In pre-Saharan Tunisia, the Matmata "mountains" (a large cuesta with its eastern front dominating Jeffara and the Gulf of Gabes) are partly covered by a desert loess that was both thick and soft enough to have been hollowed out to create troglodyte habitats, which have built the reputation and the fortune from tourism for this region.

## 1.4. Aeolian systems

In the continuum of dune forms, we rarely see single formations, except for the barchans, as they are usually joined together. We have seen that there are a variety of combinations of sifs, even if these can be reduced to two broad types. However, ghourds present second-order formations, where star dunes come together to form a large ghourd, like those seen in Issaouane n'Tiffernine (Figure 1.34). Moreover, these dunes rarely take up all available space, with more or less sandy spaces often separating the dunes. Together, these all make up aeolian systems insofar as they are chiefly related to the interaction between the atmosphere and the topography, including the dunes themselves. The systems are varied, however it is possible to identify certain types. We will start with simple combinations of dunes, building them into increasingly complex systems that often end up constructing ergs or systems that are even more astonishing, related to wind action but also other atmospheric processes, especially wave processes.

### 1.4.1. *Initial combinations: arrow, elb and draa*

#### 1.4.1.1. *Sand arrow*

**A sand arrow** is the simplest combination which results from the interaction between the relief and the transport of material by the wind. These arrows occur frequently in regions with a prevailing wind, like in the meridional Sahara, where the Harmattan blows, and they are sometimes aligned along the direction of the resulting effective local winds (Figure 1.37).

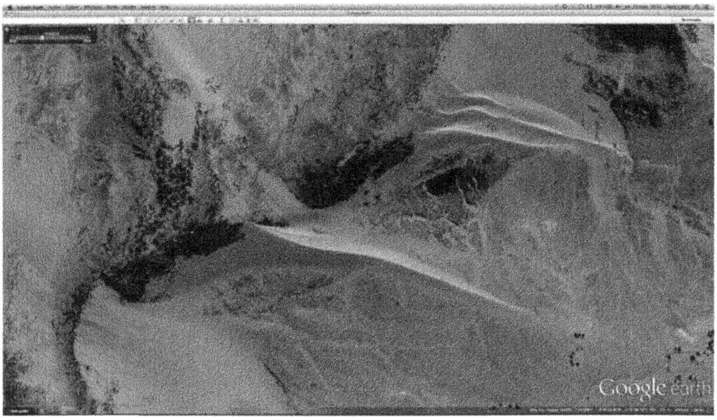

**Figure 1.37.** *Sand arrows leeward of the Fachi "cliff" on the meridional border of Ténéré (Niger) (source: Google Earth). For a color version of this figure, see www.iste.co.uk/bourrie/deserts3.zip*

#### 1.4.1.2. *Elb*

**An elb** (Figure 1.38) is the combination that appears to be the simplest. Its complexity is due to the presence of two types of dunes that play different functions in the same structure. Just as barchans are the largest possible type of sifs, the elb is the largest possible evolution of a silk. It has a fixed area and is elongated (sometimes a single silk, sometimes several silks aligned or in a relay), where on the windward side there is a zone of accumulation of thick nappages and/or dunes without sharp edges. Together this forms an oblique line of a stock of sand, which produces barchans. These allow the "exportation" of the sand in the form of a stock of mobile sand (see above), but again, an unknown proportion – probably quite large – leaves the elb in the form of free sand. This can be seen on satellite images as clear tracks that are parallel to the axis of the barchans. The elbs may grow to 10 km, or even tens of kilometers, sometimes through the elbs being lined up. Why these structures formed, and the reasons for their quasi-stability (aerial photographs of the IGN dating back half a century to 1957–1958 show no great difference in their position) remained unknown until recently (see below). Despite its apparent simplicity, the elb poses some problems, already examined earlier when studying silks, chiefly its stable position and its high obliqueness with respect to the resulting effective winds, expressed by the axis of the barchans, as demonstrated in the text where we studied these dunes.

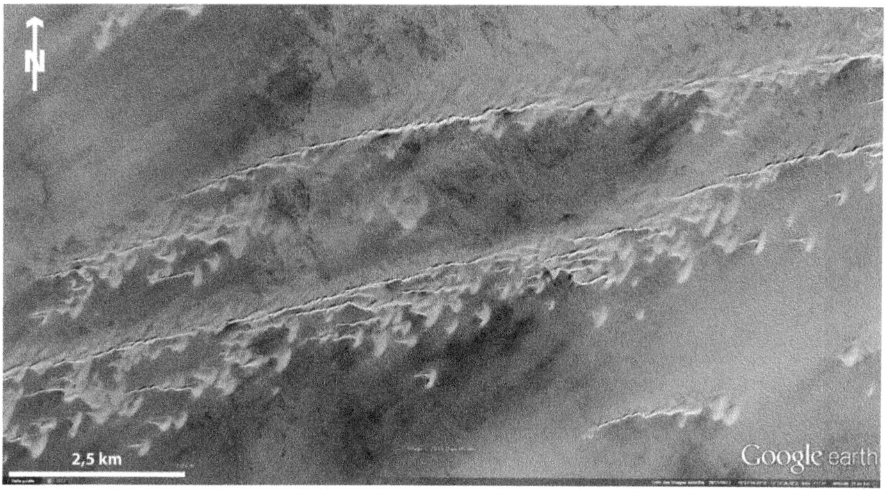


**Figure 1.38.** *Two dextrous elbs about a 100 km to the NW of Dirkou (Niger) (source: Google Earth). The most meridional of these, which is only partly seen, is about* 30 km *long. Many barchans emerge from this, showing the great divergence between the elb and the resulting effective winds. The septentrional elb, which is even more divergent, blocks a significant accumulation of the sand on the windward side, but does not give rise to many barchans. For a color version of this figure, see www.iste.co.uk/bourrie/deserts3.zip*

#### 1.4.1.3. *Draa*

**A draa** (*arm*, in Arabic) is an alignment of ghourds (Figure 1.39) that can stretch out over tens of kilometers in length in one particular direction that has no clear relationship with the direction on the wind, as they are found in regions without any prevailing wind. Because of its height, it may sometimes be covered with aklés in between the ghourds. It then forms more complex formations that we will study later.

(a) Chains of ghourds forming a draa, separated by gassis (*see below*) in the central part of the Grand Erg Oriental.

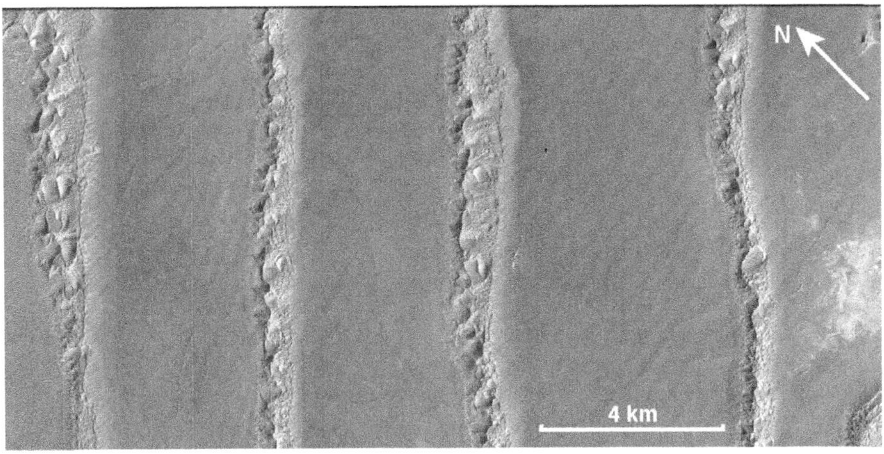


(b) Draas separated by large gassis in the Erg Chech.

**Figure 1.39.** *Draas in Algeria (source: Google Earth)*

### 1.4.2. *Interdune spaces*

Draas are the clearest example of the "default" existence of an important component in the organization of dunes: the space separating them. Indeed, it is quite rare that dunes are completely joined together. The areas that separate the dunes are of two types: sandy spaces and spaces that are completely lacking in any mobile sandy material.

The most frequent depression, especially in the ghourds, is the **sahane** (Figure 1.40), defined by Gallois (1965) as being "an almost flat stretch surrounded by ghourds". It may be covered in sand, especially when it cannot reach the substrate, sometimes covered by coarse sands, by a serir or reg, and surrounded by ghourds. Its shape varies depending on how the sifs and ghourds around it are organized. Sometimes it is very irregular, when it is formed by the division of silks with several structures, sometimes more geometric (Figure 1.40). It attains its maximal size when it is formed by the cutting up of an alignment of structures, like draas (see above).

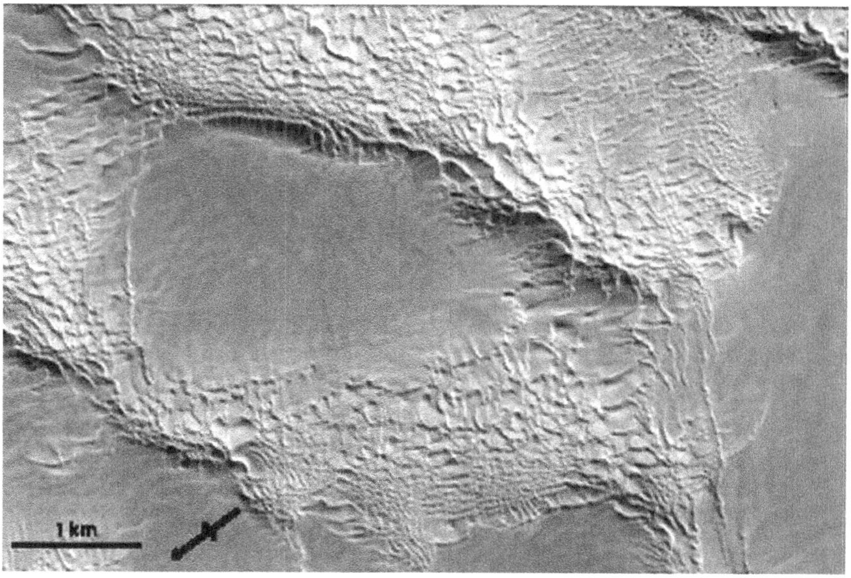


**Figure 1.40.** *A sahane surrounded by ghourds, silks and aklés in Edeyen d'Oubari (Libya) (source: Google Earth)*

However, interdune spaces are often quite rectilinear between complex dune formations that are also rectilinear. In this case, a specific terminology has been used.

**A feidj** is a sandy, interdune corridor. It may be lacking in dunes, but they are sometimes simply less dense here than in the rest of the erg. The thickness of sand, which is often quite coarse, is often hard to estimate, but may be quite thick (Figure 1.41); it is often organized into accumulations with a gentle slope, such as the zibars described earlier (Figure 1.22).

**Gassis** are rectilinear interdune spaces, which may sometimes be very long (measuring 10 km or even 100 km), separating the draas and chains of ghourds of a similar length (Figure 1.31). Unlike feidjs, their surface, which is lacking in fine or medium sand, is chiefly made up of very coarse sand, gravel or serirs. The regularity in the alternation between areas with dunes and the gassis must be explored and will be discussed later.

Gassis and feidjs are very important for human activity in deserts as they are the only paths for easy passage in ergs, for instance in the Erg of Fachi-Bilma (Figure 1.22(b)). Some of them even constitute major axes: the *Gassi Touil* (the "Large Gassi") in the Grand Erg Oriental, is the major axis for traffic in the Eastern Algerian Sahara. It makes it possible to travel through this erg from north to south, without having to cross over a single dune.

### 1.4.3. *Polygenic formations*

An analytical approach does not allow us to take the variety of combinations of constituent forms in superficial aeolian formations into account. Large dune structures, for instance, combine forms of accumulation, ablation, and coarse and fine sands, all pointing to a complex geomorphological history. We illustrate this through a few cases presented in increasing order of complexity.

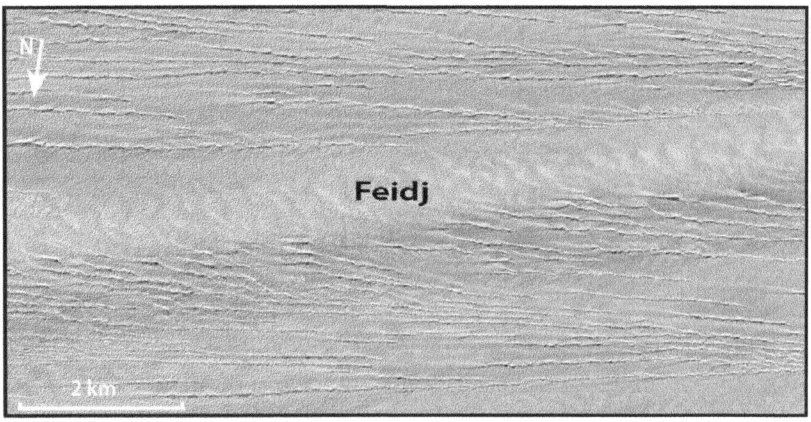


**Figure 1.41.** *Feidj in the Erg of Fachi-Bilma (Niger), almost entirely covered in zibars (see above), surrounded by undulations covered in silks*

#### 1.4.3.1. *Renewed erosion in dune structures*

A new wave of erosion in a dune is the simplest form of polygenic functioning as it corresponds to an alternation of processes. However, this must be examined in a nuanced manner. Certain alternations may be syngenetic, i.e. they may be part of the processes that form the dune. The best example of this is the case of sifs and silks, which can be seen in Figure 1.23. More generally speaking, all of the stratifications that are interwoven through the dunes are of this order and they correspond to variations in the direction and force of the wind. The variations may also be quasi-instantaneous – a sandstorm – or seasonal.

Renewed erosion *stricto sensu* is the partial, or sometimes total, destruction of sandy accumulations related to a renewal of the erosive capacity of the wind. For example, a decrease in available sand could provoke this, when an increase in green cover partly fixes the sand and increases available wind energy. On the contrary, any climatic or anthropic deterioration will destroy the continuous green cover over the fixed dunes, opening up breaches that allow the movement of sand. This was often the case in the Sahel, in dunes in the *Ogolian* erg, established during the last glacial maximum, around 18,000 BP (Rognon 1994), and was then completely fixed during the Holocene humid period, around 12,000 BP to 5000 BP. The aeolian dynamic was reactivated toward the end of the 20th century (Figure 1.42) due to the severe droughts that affected the entire Sahel region in the 1970s and 1980s and whose effects were exacerbated by extensive anthropic pressure, with large amounts of soil being carried away and redeposited in the form of active dunes.

In all cases, there are the formation of furrows oriented in the direction of the resulting effective winds, as they are created through ablation, without any possibility of oblique deposits.

#### 1.4.3.2. *Sand Ridges*

These formations were studied and defined in Australia. Some authors no longer use this term (Thomas 1997), preferring *linear dunes* instead, even though the former term appears more precise and limited in a semi-arid domain. It designates dunes that are almost rectilinear, with sharp edges, covering a small area and separated by fixed spaces (Figure 1.43).

**Figure 1.42.** *Renewed erosion in dunes in the Ogolian erg (Mali). These fixed dunes are again set into motion with the creation of depressions that eat away at the older dune, a remnant of which is all that remains at the center. For a color version of this figure, see www.iste.co.uk/bourrie/deserts3.zip*

These ridges have a precise ratio between their height and the gap between the crests, which varies from erg to erg, and often have Y-junctions or even X-shaped junctions (Figure 1.43(c)), which disproves their longitudinal character. Dunes in the Simpson Desert (Figure 1.44) were described by Twidale (1980). They can be up to 200 km long. If the space between the dunes measures about 200 m, their height will be about 15 m. However, if the space between them is between 500 m to 1000 m, the height may be as much as 35 m. If the crests have a sharp, asymmetric form, the intermediate areas are more complex, semi-fixed or fixed by vegetation and/or granulometry. The simplicity of the shape masks what can be a very complex genesis, with granulometry that is quite coarse, the median size of the grains increasing from 300 μm on the slopes to 500 μm on the crest, strangely similar to the granulometry seen in megaripples.

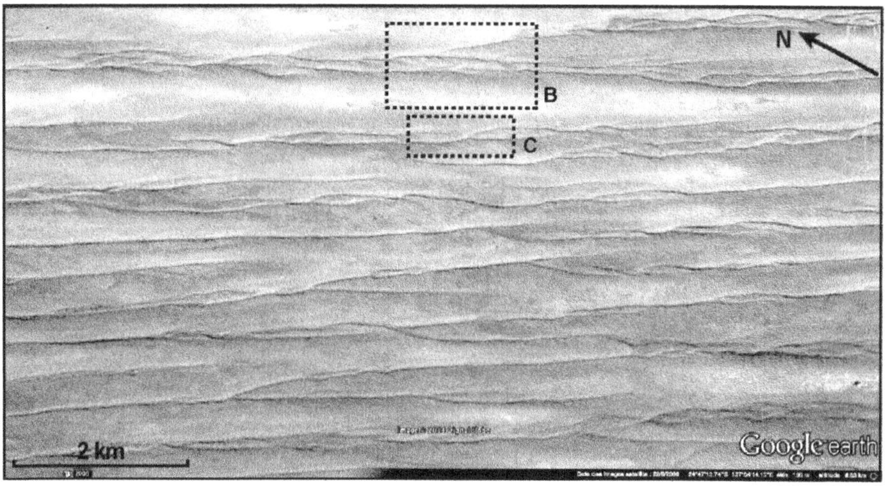


(a) A group of sand ridges, where the crests of active sands diverge and then converge (scale 2,000 m).

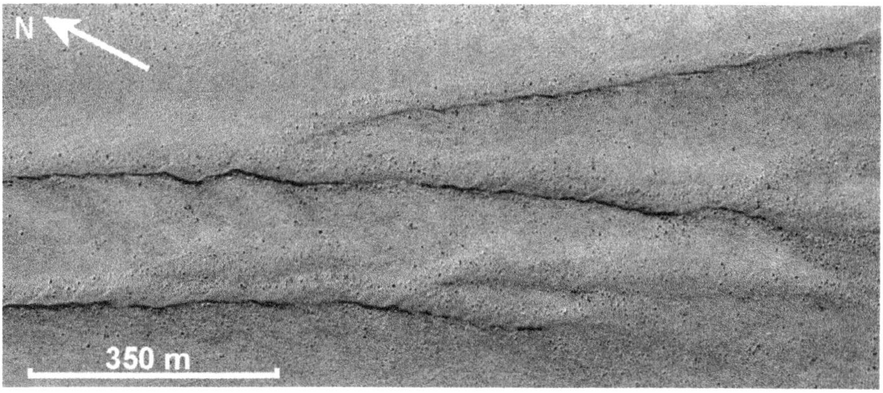


(b) A point of divergence for the crests.

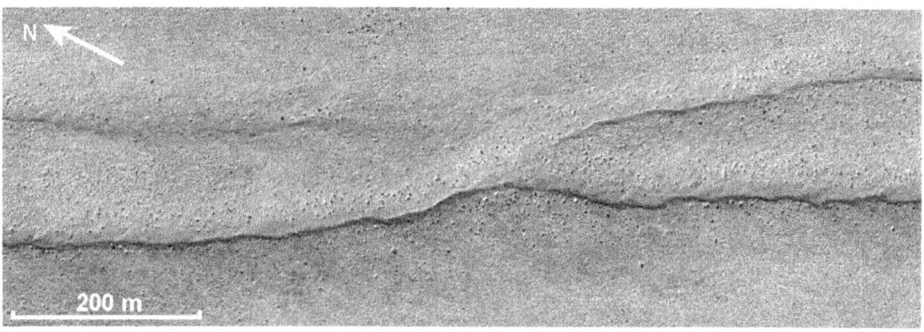


(c) A point of intersection of the crests.

**Figure 1.43.** *Sand ridges in the Simpson desert 140 km west of Birdsville (source: Google Earth)*

#### 1.4.3.3. *Polygenic systems in regions with prevailing winds*

In these regions, polygenic systems are often characterized by the opposing arrangement of forms with different regions, resulting from a series of processes that resulted from climatic modifications. The southern Sahara and the Sahel region, where the harmattan blows and the climatic variations of the recent Quaternary were very pronounced, are regions where these systems cover large areas from the Atlantic to Sudan. In the Erg of Fachi-Bilma, Chad and Niger (Figure 1.41), a "stationary erg" forms large undulations, which may stretch out up to 100 km in length, separated by feidjs. It has coarse granulometry, especially on the surface, where a deflation pavement leads to a concentration of grains larger than 630 μm, which are much rarer in sand taken at the same location, but at a depth of 5 – 10 cm. This erg has silks with active sands (which are much finer) superimposed on it. These silks have an overall deviation toward the right resulting in a large angular deviation between the large undulations and the functional dunes. It is not very probable that these two kinds of forms were created simultaneously, as that would require ablation and deposit to occur together. It is more logical to interpret this landscape as having been built up over different periods in time: the formation of an ancient erg, made up of very large dunes, with a generally complex history; in this case, this is probably the extension of the Ogolian erg (see above) around 18,000 years BP. This was then followed by a period of erosion with fixation, associated with the humid Holocene period, with the formation oriented in the direction of the resultant wind direction. Finally, since the return of aridity, active, divergent forms have appeared. However, this sequence is probably only the most recent of a series of earlier alternations that accompanied the Quaternary climatic variations.

#### 1.4.3.4. *Polygenic systems in regions with variable winds: dunes in the NE region of the Grand Erg Occidental*

In the septentrional Sahara, there are many complex forms, however they are often confusing and it is difficult to accurately describe their arrangement. In general, the wind does not have a clearly dominant direction, so it is not possible for dunes to form aligned with the prevailing winds.

The large collection of dunes in the northeastern part of the Grand Erg Occidental (Callot 1991) is a representative example of this kind of formation (Figure 1.45).

**Figure 1.44.** *Sand ridge on the eastern border of the Simpson Desert, 40 km from Birdsville (source: commons.wikimedia, photo by Paul Hanly, September 10, 2007). For a color version of this figure, see www.iste.co.uk/bourrie/deserts3.zip*

The dunes, about 100 m high, are chiefly made up of large domes of coarse sand (median of 700 μm), surmounted by structures with active sand (median of 220 μm). These structures are arranged into groups aligned in one principal direction (NW-SE). A secondary NE-SW alignment disrupts the regularity of these lines.

The corridors between the domes are in the form of feidjs with a substratum of sand that is coarse on the surface, on a layer whose thickness ranges from 10 m to tens of meters. The occurrence of deflation is heightened here in the "cauldrons" that perforate the feidjs, giving them an irregular topography, but does not reach the substrate. Lacustrine deposits are found in the "cauldrons", formed during the height of the Holocene humid period, between 10,000 and 7,000 years BP (Callot and Fontugne 2008). There are also other more advanced deposits (and thus older), which can be attributed to the Upper Pleistocene, or even the Middle Pleistocene. These show that the structure of the dunes is remarkably stable, since the deflation areas remain in the same position. This has also given the position of the dunes equal stability. These structures, which were established during a more humid period, are reactivated in the same place. They are presently made up (Figure 1.46a) of a superimposition of aklés and ghourds with the granulometry of active sands over structures with coarser sands (Figure 1.46b).

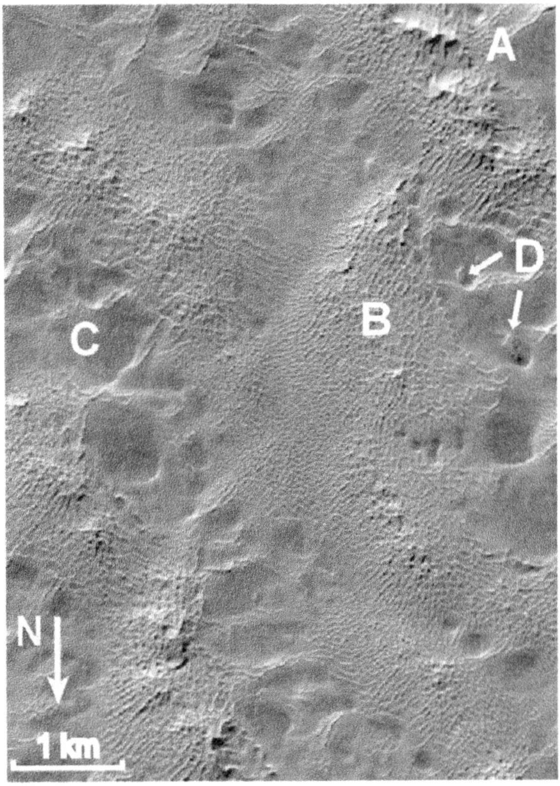


**Figure 1.45.** *Arrangements of dunes in the northeastern part of the Grand Erg Occidental. The ghourds (A) and the aklés (B) cover the ancient dune structures that separate feidjs perforated by deflation cauldrons (C), in which outcrops of lacustrine deposits can be seen (D) (source: Spot Panchromatique)*

The two types of sands and dunes are identical to those described in the Erg of Fachi-Bilma, discussed earlier. However, the forms here are completely superimposed on one another, while in the Erg of Fachi-Bilma they were slightly offset, allowing more of the older forms with coarse sands to be seen.

It is almost impossible to explain dune systems and the interdune spaces that are so complex over a scale of a kilometer, when we only see the arrangement of a few structures that are in proximity to each other. Only a more "removed" approach, on a regional scale, makes it possible to understand the organization of these systems and, sometimes, to understand the causes that led to their formation.

(a) Ground view of the large dunes of the Erg

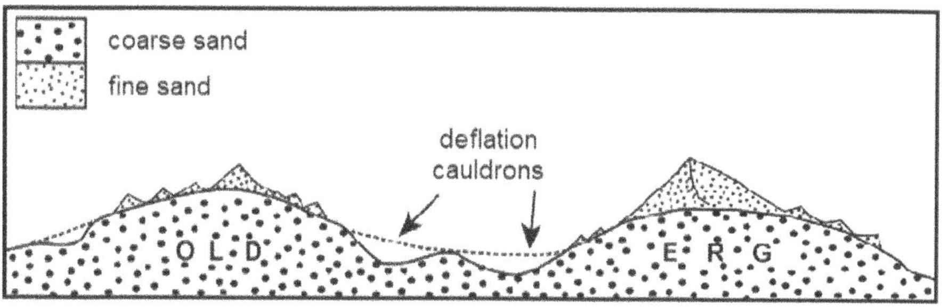


(b) A cross-section of the large dunes

**Figure □.□□.** *Ground view and cross-section of the large dunes of the erg. For a color version of this figure, see www.iste.co.uk/ bourrie/ deserts3.zip*

## 1.5. Ergs, the most complex aeolian systems

Ergs represent the most complex aeolian systems. They offer such vast scope for study that we can only offer a brief summary here, however that is enough to show how ergs can sometimes constrain human activity.

### 1.5.1. *Definition*

The term "erg" comes from Arabic and has become the most widely used term, however there are other terms that are also used: *sand sea* in English, *Edeyen* in Libya, *Issaouane* in the Eastern Algerian Sahara, *Goz* in the southeast Sahara and *Koum* in Central Asia. Given the variety in dune organizations, authors diverge greatly on how to define them. The definitions proposed are often too restrictive and prevent certain large ergs from being considered as large ergs. For example, Cooke, Warren, and Goudie (1993) restrict the term "sand sea" to any area measuring at least 30,000 $km^2$, where discontinuities in the sandy cover are not more widespread than the dunes. Below this, they use the term *dunefield*, starting from 10 dunes, with the same restrictions on the area covered by the dunes. Wilson (1971) has a less restrictive term, *active erg*, for an area of at least 125 $km^2$, where at least 20% of the surface is covered with active sand.

The nested scales and knowledge brought in by satellite imaging makes it possible to propose a broader definition:

*An erg is a widespread concentration of sandy structures, joined together or juxtaposed, active or partially fixed, stretching over more than 100 $km^2$.*

While being close to the definition proposed by Wilson (1971), it makes no claim to perfection: the criteria for the area covered, especially, is somewhat arbitrary, being more practical than based on any precise break in the continuum of dune-covered areas. Unlike most definitions, however, this does not include criteria related to the percentage of the surface that must be covered by dunes. Indeed, some large ergs, especially the Chech and Iguidi (Algeria–Mali), are largely vast gassis (Figure 1.39(b)) separated by draas that sometimes cover less than 20% of the surface. Being impassable over tens of kilometers, they make the region a true erg, almost impossible to cross perpendicular to the gassis.

### 1.5.2. *The formation of ergs*

Several parameters may come into play in the formation of an erg and its origins are often hypothetical: the inflow of sand is essential, but the force and direction of the wind, the duration for which it blows and the topographic context on a regional level are all important factors. For instance, large basins surrounded by mountains are favorable areas if they are not submerged by water. The Takla Makan is one of the most perfect examples of this. Sometimes there are other contributing factors that can be clearly seen: for instance, in Afghanistan, the "red" erg in the Registan, close to Kandahar, is notably bordered in the north by the Helmand, Arghandab and the Arghistan (Figure 1.47), which would indicate that it is fed by alluvial deposits from these water bodies. In large ergs, there may sometimes be very complex causes: dynamics and the age of the dunes may vary completely from one part of the erg to another.

#### 1.5.2.1. *Localization and extension*

Ergs are distributed in varied manners across different deserts and do not cover a significant portion of their area, especially if we consider their discontinuous nature. Wilson (1973) states that large ergs measuring over 30,000 km$^2$ (of which there are not many) represent the majority of area covered by dunes on the earth's surface (around 85%). The vastest of these is the Rub'Al Khali, in the Arabian peninsula, whose area is equal to that of France. Note that 45% of the area covered by ergs is in Asia, 34% in Africa and 20% in Australia (Mainguet and Callot 1978).

#### 1.5.2.2. *Typology*

These definitions clearly show that ergs are not necessarily continuous sandy areas. In fact, they are highly varied, which makes it possible to classify them in different ways.

**Based on how much they have been fixed by plants:** This criterion is an essential one for human activity. Fixed ergs, also called "covered" ergs can be found in semi-deserts or on the periphery of real deserts. They are often conducive to nomadic livestock rearing, because of their capacity to absorb precipitation through the sand and the presence of a loose substrate. On the other hand, they are also very sensitive to renewed wind erosion, as they are made up of material brought in by the wind.

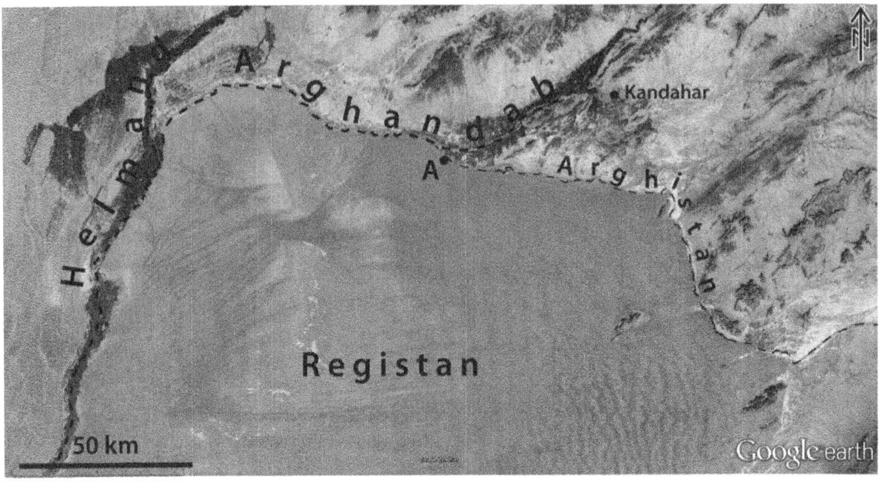


(a) Overall view.

(b) Detailed view of the contact between the water body and the dunes at point A; the cultivated land on the right bank offers a stark contrast to the active and unbroken sand on the left bank.

**Figure 1.47.** *Northern part of the erg in the Registan, clearly limited by the waterbody that supplies it with material (source: Google Earth). For a color version of this figure, see www.iste.co.uk/bourrie/deserts3.zip*

**Based on the dunes that make up the erg**, or form the major part of the erg: ergs made up of silks, frequently seen in the Southern Sahara, and ergs made up of ghourds, frequently seen in the Northern Sahara, given the absence of any markedly prevailing winds. However, the majority of ergs, especially the large ones, are complex and formed of several different types of structures (silks and ghourds, or silks and barchans, for example).

**Based on the proportion of sand cover**, already seen in the definition of an erg.

While "continuous" ergs exist, in almost all cases a part of their surface is made up of coarse sand that is not very mobile, forming zibars. The entire central and western sections of the Erg of Fachi-Bilma in Niger (Figure 1.41) is a remarkable example of this. The active dunes are not, thus, joined together, and are separated by feidjs. There are significant differences in the proportion of active and fixed sands. In the Great Sand Sea in Libya, active dunes are in the minority. On the other hand, in the northeastern part of the Grand Erg Occidental in Algeria, which we studied earlier (Figures 1.45 and 1.46), the areas with active dunes predominate over others.

Ergs that are almost entirely made up of active dunes are quite rare: certain parts of Edeyen Oubari in Libya, the meridional part of the Grand Erg Occidental. Being almost impossible to cross they act as clear barriers to human activity, and remain so in ergs where active dunes are very close together, as seen in Figure 1.48. However, this figure also shows that modern technology makes it possible to construct paths (a large track, here) in environments that are just as difficult.

**But most ergs are discontinuous.** The proportion of area covered by sand is more or less variable, especially with respect to active sand. The dune groups are then separated by gassis, seen earlier, whose surface state may be a reg or a serir.

Groups of dunes may completely lack any orientation. However, satellite imaging shows that at a certain scale, structures start emerging: alignments in draas, separated by gassis (Erg Chech and Erg Iguidi in the Southwestern Algerian Sahara and in Mali; the western part of the Grand Erg Oriental); quadrillages (the septentrionale part of the Erg Iguidi). The organization of large ergs is seen to vary in different sectors (see Figures 1.49 and 1.50). The Edeyen of Mourzouk, long cited as one of the most continuous ergs on earth, belongs to this category. Satellite images show a large part dominated by coarse sand (probably zizars) with bouquets of silks, and a southeastern part where large ghourds surround the sahanes. Indeed, it is the difficulties encountered in crossing this erg, because of the total absence of any wells, that has lent it its fearsome reputation.

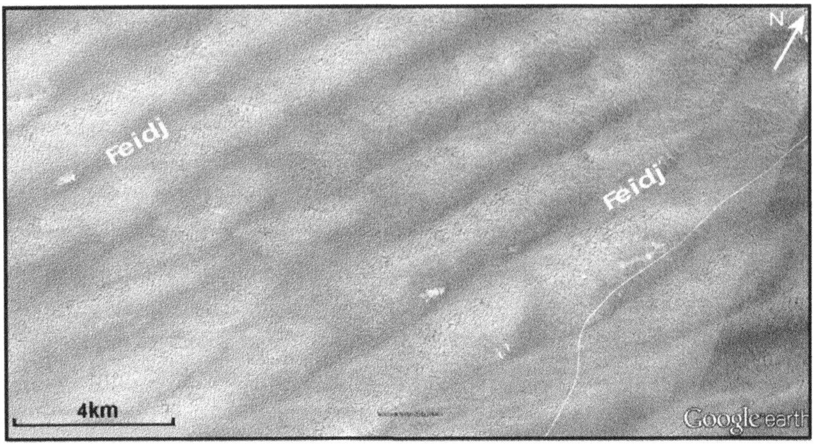


(a) Scale: 4 km.

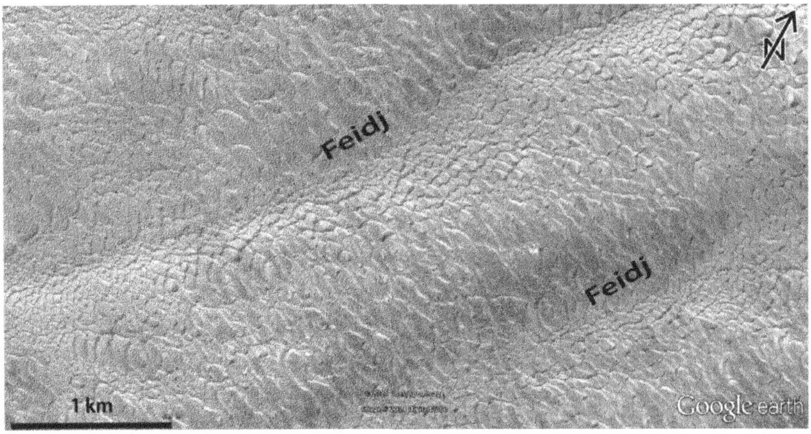


(b) Scale: 1 km.

**Figure 1.48.** *Arrangement of dunes in the southern part of the central section of the Grand Erg Occidental (Algeria) (source: Google Earth). Large undulations are separated by feidjs that can barely be recognized under the continuous cover of an aklé*

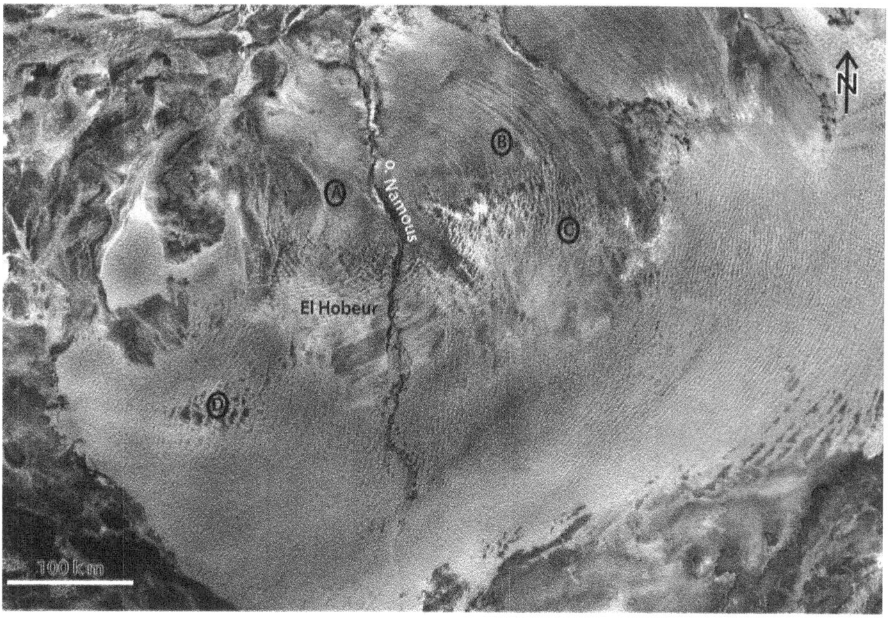


**Figure 1.49.** *The central and septentrional part of the Grand Erg Occidental (source: Google Earth). A semi-circular structure (toward the middle, marked A) commands the complete organization of the dunes in the central and eastern part of the erg, as well as those in the hamadas located to the north of the erg. A N-S to NNE-SSW structure cut through the previous one on either side of the Oued Namous. For a color version of this figure, see www.iste.co.uk/bourrie/deserts3.zip*

### 1.5.3. *From the stability of forms... to a whole new scale: the Grand Erg Occidental*

Because of their dimensions, the largest ergs often cannot be categorized, except by highlighting their chief unique feature. They are made up of parts that are dense, and others that are more "open", without any real relationship between the density and the central position of that section, with the densest zones often being on the periphery. The types of dunes vary in the same manner. The Grand Erg Occidental, which does not extend very far (about 80,000 km$^2$), is highly complex and presents most particular features of large ergs, at least those seen in the Sahara.

In this erg (Callot 1991), the sand is brought in by winds picking up alluvial deposits brought in during each flood by the water bodies coming in from the Saharan Atlas.

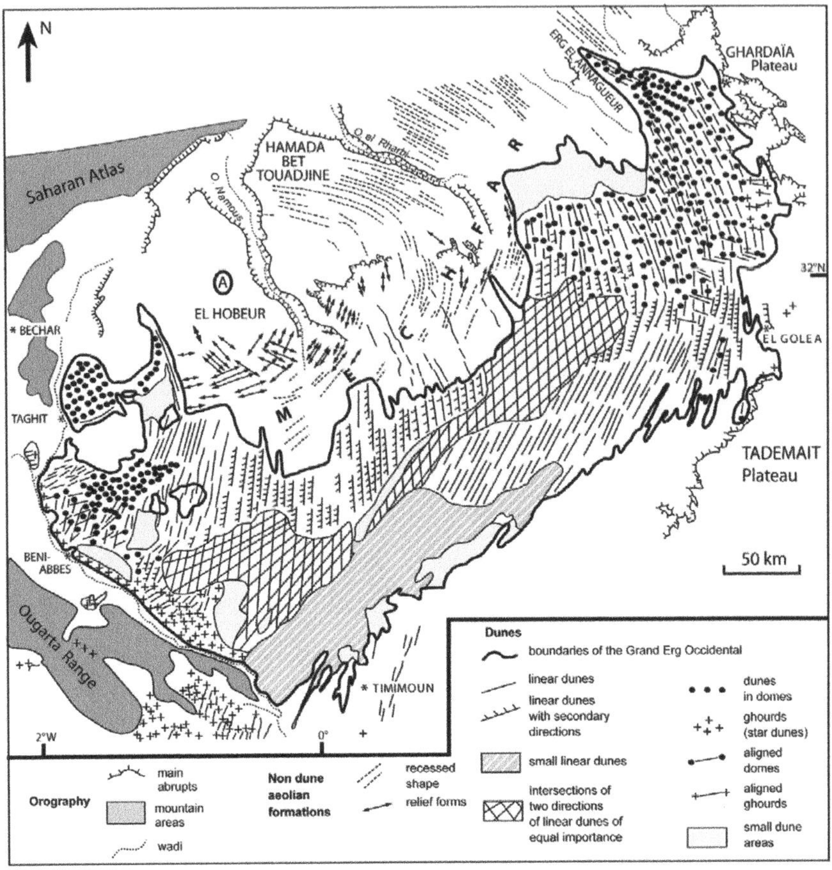


**Figure 1.50.** *Structure of the Grand Erg Occidental (as per Callot 2008, modified)*

There appears to be a clear opposition between the two parts (Figures 1.49 and 1.50):

– a central and eastern part with a markedly aeolian structure forms a semi-circle whose diameter measures several hundred kilometers and whose center is located in the central part of the Oued Namous (marked "A" in Figures 1.49 and 1.50). While dunes form predominately toward the south, the central section contains vast hydro-aeolian depressions whose formation took place over long periods (marked as C in Figure 1.49);

– a western part, where the dominant orientations are almost N-S and where there are no forms of hydro-aeolian hollowing out to be clearly seen at D. Consequently, it is assumed to be of a more recent age. The two parts are largely superimposed in the central section of the erg.

To the NW of the erg, in the El Hobeur region, there are remarkable hydero-aeolian depressions, described in section 1.2.3: (Figure 1.14): narrow rectilinear plateaus aligned NNE-SSW, which are the highest components, cut across identical structures oriented WNW-ESE that are not as high. The orientation of the strips of these lower plateaus is very "local" and cannot be seen elsewhere. These two sets of crossed plateaus, one above the other, surround closed, hydro-aeolian depressions located much lower, which are almost square (we saw these earlier). The infrequent areas with dunes are found on the highest plateaus. This topography, divided into two by the Oued Namous, is aeolian in origin, as it was formed by carrying away material from the level of the highest plateaus, which are almost entirely horizontal, and is made up of non-soluble sediments. Thus, the removal of material must date back to ancient times.

This remarkable structure, which is not very sprawling, may partly be explained by its position in the overall structure, as shown in Figures 1.49 and 1.50. Indeed, the very narrow strips of the NE-SW plateaus are located in the alignment of the circular structure of the Erg Occidental, located too close to its center, so that they turn quite quickly from the north to the south of El Hobeur (Figure 1.14). Their position as the highest entities indicates that they belong to the first structure that organized the aligned arrangement of the dune massifs in this erg.

This observation corroborates those carried out further NE, which demonstrated the ancient nature of the structure (Callot 1987; Callot 1988). This part of the structure, very clearly visible on satellite images, is not made up of dunes at all. It is found on the hamadas located between the Saharan Atlas to the north and the erg to the south (point B in Figure 1.49). The curves are those of the valleys that formed between the dune alignments during an older phase where the erg extended to the north. When the dunes disappeared, the valleys, which were barely carved upstream – being almost indiscernible on the ground – were fossilized by a Plio-Pleistocene hamadian pavement. The valleys were more deeply carved downstream, opening into large, long, closed depressions, while elongated dune massifs appear on the trips of the plateaus that separate the depressions. Further south, dune massifs extend the curve and become parallel to the set of feidjs in the meridional part of the erg. This continuity between the fossilized forms in the north and the active, current formations in the south show that this aeolian structure has not moved since formation.

The origin of this structure, the fact that it is divided by another N-S structure to the west of the erg, and by a third at El Hobeur, remains, enigmatic. This astonishing region has been identified and described using aerial photographs by Clos-Arceduc (1965), who was unable to visit it on the ground. However, he was able to show the alignment of valleys, the depressions and the dunes at El Hobeur. He thus proposed a theory to explain this, which agrees with what can be seen on the ground. He interprets the position of the dune as resulting from the interference of stationary atmospheric waves. In the antinodes of the waves, the vertical agitation of the atmosphere prevents sand being deposited and promotes aeolian erosion. On the contrary, in the nodes of the wave, the absence of any agitation is conducive to the accumulation of sand, which protects the surface against erosion. The observations of the curved structure in the zone where these indented "valleys" transform into depressions (Figure 1.51) corroborates this interpretation: throughout this part of the erg, dune fields are largely found on the plateaus that are, theoretically, swept by the wind, while being almost absent in the valleys, which one would think are more favorable to accumulation, as they are more sheltered.

The remarkable observations and interpretations made by Clos-Arceduc (1965), borne out of many examples, may be very convincing, but cannot be proven. However, they have opened up several interesting avenues to explore: how can we explain the stability of structures generated by these oscillations that are stationary on a geological scale? A stability that is necessary for aeolian or hydro-aeolian ablation to create entire landscapes starting from few cubic kilometers of material available. How and why do new oscillation systems appear, like those to the west of the Grand Erg Occidental, or again in the El Hobeur region, described earlier? All of this must be studied in detail. The only hypothesis we can propose for now is very geographic in nature: once the system is in place, the topography maintains the system of stationary waves.

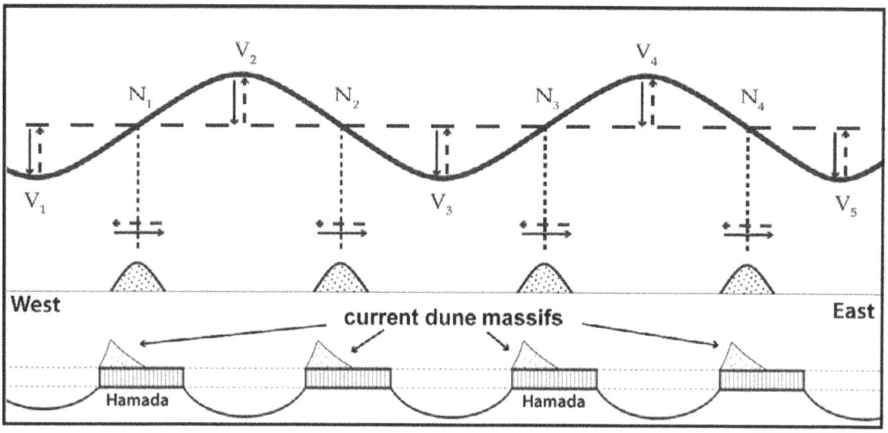


**Figure 1.51.** *Relationship between stationary oscillations and relief features along the north-central border of the Grand Erg Occidental (as per Clos-Arceduc 1965; Venard 2012, modified)*

It must be noted that this explanation was also proposed by Clos-Arceduc (1965) to explain the alignments of the draas, which we saw earlier. In this case, there is no significant erosion in the gassis, but accumulation that chiefly occurs at the nodal points. It is also an effective means of explaining all the complex geometric arrangements of the ergs: triangular, lozenge-shaped, rectangular etc., described in the preceding sections.

The Grand Erg Occidental is one of the most complex cases in the world. It is thus fitting that it forms the conclusion of this study of the aeolian form. This is all the more appropriate since the "Grand Erg Occidental system" includes almost all processes described in this section on aeolian processes. Indeed, it includes large forms of aeolian hollowing; some of its structures are "fossilized" under the surface of a hamada; ancient actions combine with present-day actions; it is one of those rare ergs where certain sections are completely covered in dunes and it is almost impossible to travel through some of its sections. Studying this erg also shows us how many aspects of aeolian processes and formations remain unexplained today, even though a few have been recently explained.

## 1.6. Conclusion: a new paradigm to explain the organization and orientation of active dunes

At multiple points in this study on accumulation forms, especially those with active sands, we have described dunes that seem to contradict certain features of the environment in which they are located, especially the resulting winds. This is particularly the case for large, elongated and rectilinear dunes, like the silks, for which we have proposed an explanation related to an interaction between the "dune object" and the wind. What was once only a hypothesis was recently the subject of a physical demonstration that offered researchers a new paradigm to understand aeolian formations. This was obtained through modeling (Zhang et al. 2012). Subsequent experimentation (Courrech-du-Pont, Narteau, and Gao 2014) made it possible to recreate, on a smaller scale, the exact shapes of the barchans, silks and elbs observed in the Erg of Fachi-Bilma (for example, Figure 1.38). The experiment was carried out by circulating containers of a material with a precise granulometry (90 μm) in an elongated basin filled with water; the angles of rotation of the container made it possible to simulate different "winds". The container would move fast in one direction, to simulate wind, and slowly in another so that there would be no displacement effect. Finally, an experiment on a "natural scale" was carried out in the deserts of China (Ping et al. 2014) by bulldozing 16 hectares of dunes and then observing their reconstruction while measuring wind flux.

This model takes several parameters into account. First, it considers that there is a principal wind and a secondary, weaker wind – the ratio in the case that was presented was 1/5 – which may result in variable angles, but yields forms that are clearly defined in the case presented, with an obtuse angle (130°). Second, the model varied another parameter, namely the supply of sand.

Figure 1.52 depicts the results of the three simulations carried out in conditions that were strictly identical apart from the quantity of sand available:

– in A, there is a continuous layer of sand at the bottom of the container and no fresh sand is added during the rest of the experiment;

– in B, the initial conditions are identical, but there is a constant addition of sand in the "interdune" spaces to simulate conditions of an unlimited supply of sand;

– in C, the container has no sand, but at one point there is a conical deposit of 1 g of sediment, which is renewed by 0.4 g every 20 circulation cycles.

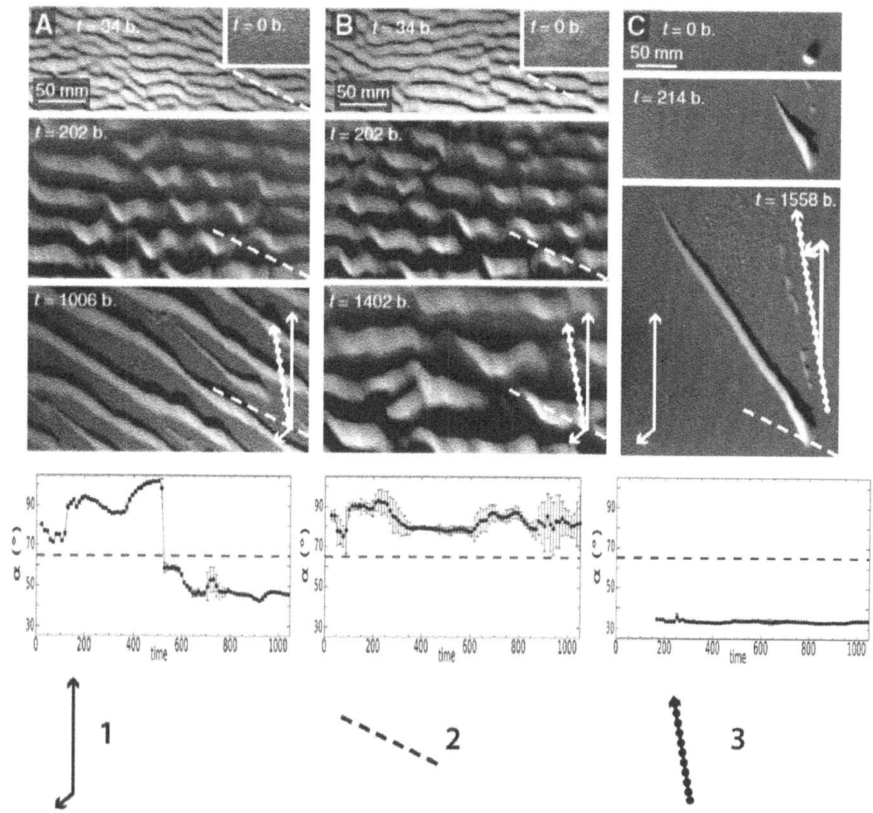


**Figure 1.52.** *Simulation of the differences in the creation of the formations in conditions where the circulation of fluid is identical based on the quantity of sand available to the fluid (as per Courrech du Pont et al., 2014, modified). 1: direction of the two flows; 2: bisectrix of the angle made by both directions; 3: vectorial resultant of the two flows*

The formations obtained in strictly identical "wind" conditions are radically different from one another, demonstrating that it is the supply of sand that governs the orientation of the ripples: A and B initially grow in the same manner, forming structures that are roughly transverse as long as there is a continuous layer of sand in the container. However, in A, when discontinuities arise because the sand stock is not replenished, there is an immediate reorientation of the ripples, expressed by the vertical party of the graph showing the orientation of the dunes. During simulation B, there is no overall variation in orientation: the structures grow in size, with greater variability in orientations, which is probably due to the appearance of faces on the "larger" structures, which are thus more complex than simple ripples.

Simulation C demonstrates an even more astonishing arrangement: from an occasional source in an area with a large sand deficit, there appears to be small structures that circulate in the direction of the resulting winds, which are micro-barchans and much larger structures, strictly matching the arrows and silks described earlier, and presenting, in the figure, a leftward deviation due to the action of the "secondary", divergent wind.

The explanation for these deviations corresponds to the scale: the authors demonstrated that the direction of the "flow" at the level of the crest of the silks strictly matches the orientation of the dune. However, this is not the same for the direction of flux over terrain free of dunes and not saturated with sand, as the flux at the crest is strongly influenced by the presence of the dune structure. The physical demonstration does, thus, match the intuitive explanation that was unprovable earlier. It also shows – and this essential point is quite new – that the wind is not the only important variable in how dunes are oriented: the flux being "saturated" or "non-saturated" by sand provokes different organizations of dune forms with differences in the angle measuring tens of degrees! Thus, taking into account both the wind and the supply of sand in the same model led to a simple and proven explanation for the orientation of the dunes. While we showed, earlier in the discussion, that it was guessed that these structures themselves influenced their orientation – which has now been brilliantly demonstrated – the essential role played by the supply of sand in the orientation of dunes is a new and remarkable finding (Callot and Crépy 2018).

### 1.7. Martian dunes, still mobile...

In the 21st century, it would be abnormal to talk about aeolian action in arid regions without speaking of a totally arid territory that is, as yet, untouched by humans, with the exception of some man-made machines: Mars, whose topography is better known (through satellite observation) than certain areas of Earth!

Given the current absence of water and life on the planet, weathering is highly reduced. Traces of aeolian action on Mars have long-been studied. Even in the 1990s, in a volume edited by Thomas (1997) that focused on the geomorphology of arid regions, there was an entire chapter dedicated to this planet (Wells and Zimbelman 1997). A large number of publications have since focused on wind, comparing terrestrial aeolian formations, aeolian processes and even granulometry with those of other planets, especially Mars (Kok et al. 2012). Many eolian formations have been observed, for example: transverse dunes and yardangs. Figure 1.53, taken from an image with a definition of 50 cm, depicts a region where the "sand" descends from a plateau to form sandy arrows and then lines of barchans (Figure 1.54(a)) with the edifices sometimes merging and forming elongated wings. We also see megaripples as well as ripples that are very visible on the dunes (Figure 1.54).

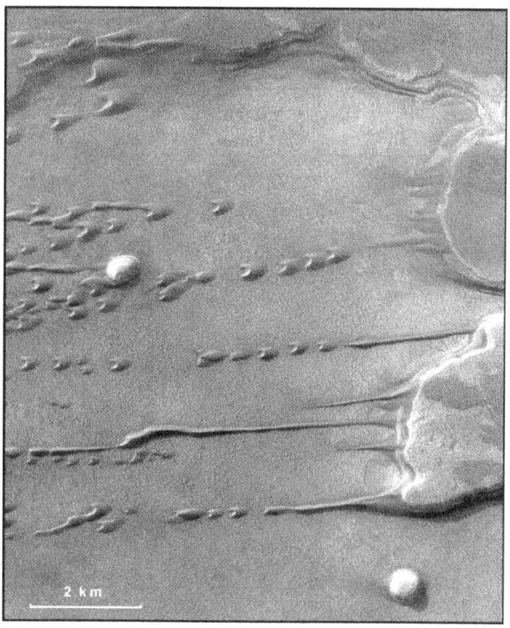


**Figure 1.53.** *Dunes leeward of a Martian plateau in the Hellespont region (latitude: −41,407°, longitude: 44,587°, Photo credit NASA/JPL/University of Arizona). Overall view (about 10 x 8.5 km). Sandy arrows continuing as lines of barchans, sometimes grouped together into silks. At the foot of the plateau, we have megaripples*

In more recent missions, the images obtained by observation satellites reveal remarkable barchans on which indicators of current activity can be seen (Silvestro et al. 2010) through the displacement of ripple marks and advances in the barchan-front. The Spirit Rover also observed dust devils. The intensity with which mineral aerosols were lifted is such that dust storms may mask the planet for long periods, up to a week. Dustfall is one of the problems that Mars rovers must deal with as the dust is deposited on their solar panels and disrupts their functioning.

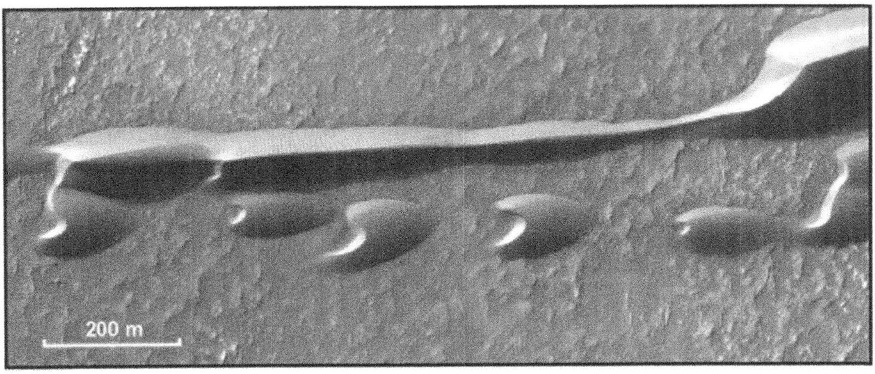


(a) Detail of the previous view (1.25 x 0.5 km).

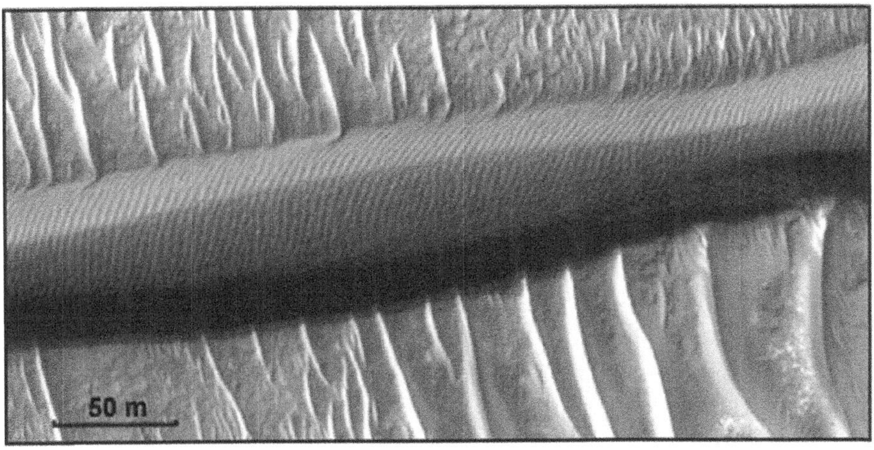


(b) Detailed view of a dune at the foot of a plateau (350 x 170 m). We can clearly see megaripples on the substrate and ripple marks on the dune.

**Figure 1.54.** *Detailed views of Martian dunes*

While the processes and dunes may seem identical to those on earth, the structures sometimes seem slightly different, perhaps because of the physical differences between the two planets: the atmospheric pressure on Mars is 150 times lower than that on Earth, while its gravity is thrice as high as Earth's gravity. The megaripples, which are very well developed, have much sharper crests – and may perhaps be dunes – and the ripple marks are larger. However, some photos taken on the ground by the Curiosity Rover (Figure 1.55) reveal landscapes that are identical to Earth's, such that without any caption, they could be mistaken for "Earthscapes", captured in the most arid regions of our globe such as the Atacama or the Egyptian-Libyan desert.

**Figure 1.55.** *Mars dune blocking a hill, photo from the Curiosity Rover (source: lemonde.fr/sciences/ portfolio/2014/08/05/les-plus-belles-photos-de-mars-prises-par-curiosity_4463930_1650684.html). For a color version of this figure, see www.iste.co.uk/bourrie/deserts3.zip*

## 1.8. References

Abolkhair, Y. (1981). "Sand encroachement by wind in Al-Hasa of Saudi Arabia." PhD. Indiana University. 196 pp.

Bagnold, R. (1960). *Physics of Blown Sand and Desert Dunes*. 3rd ed. London: Methuen. 265 pp.

Beadnell, H. (1910). "The sand dune of the Libyan desert." *Geographical Journal* (35), pp. 379–395.

Callot, Y. (1987). *Géomorphologie et paléoenvironnements de l'Atlas Saharien au Grand Erg Occidental : dynamique éolienne et paléolacs holocènes*. Mémoire des Sciences de la Terre 87-21. Paris: Université Pierre et Marie Curie. 412 pp.

— (1988). "Évolution polyphasée d'un massif dunaire subtropical : le Grand Erg Occidental (Algérie)."*Bulletin de la Société Géologique de France* 8th series, IV (6), pp. 1073–1079.

— (1991). "Le Grand Erg Occidental (Algérie) - Histoire d'un massif de dunes." *Sécheresse* 2 (1), pp. 26–39.

— (2008). "Formations éoliennes : Dynamique éolienne et formations éoliennes en domaine aride et semi-aride". *LES FORMATIONS SUPERFICIELLES, Genèse — Typologie — Classification — Paysages et environnement — Ressources et risques*. Edited by Y. Dewolf and G. Bourrié. Paris : Ellipses, p. 293-317.

Callot, Y. and M. Crépy (2018). "Le rôle de l'alimentation en sable dans la dynamique des dunes linéaires : un changement de paradigme ? Approche croisée des apports de la modélisation et des observations de terrain (Erg de Fachi-Bilma, Niger, Tchad)." *Geomorphologie : relief, processus, environnement* 24 (3), pp. 225–235.

Callot, Y. and M. Fontugne (2008). "Les sites lacustres d'âge holocène dans l'est du Grand Erg occidental (nord-ouest du Sahara algérien)." *Geomorphologie : relief, processus, environnement* (3), pp. 187–200.

Callot, Y., B. Marticorena, and G. Bergametti (2000). "Geomorphologic approach for modelling the surface features of arid environments in a model of dust emissions: Application to the Sahara desert." *Geodinamica Acta* 13 (5), pp. 245–270.

Capot-Rey, R. (1963). "Contribution à l'étude et a la présentation des barkhanes." *Travaux de l'Institut de Recherches sahariennes, Alger* (22), pp. 37–60.

Clos-Arceduc, A. (1965). "Essai d'explication des formes dunaires sahariennes." *Études de Photo-Interprétation* (4).

Cooke, R., A. Warren, and A. Goudie (1993). *Desert Geomorphology*. London: University College London Press. 526 pp.

Coudé-Gaussen, G. (1991). *Les poussières sahariennes - cycle sédimentaire et place dans les environnements et paléoenvironnements désertiques*. Paris: John Libbey Eurotext. 498 pp.

Courrech-du-Pont, S., C. Narteau, and X. Gao (2014). "Two modes for dune orientation." *Geology* 14, pp. 747–750. doi: 10.1130/G35657.1.

Coursin, A. (1964). "Observation et expériences faites en avril et mai 1956 sur les barkhanes du Souehel el Abiodh (région est de Port-Étienne)." *Bulletin de l'Institut Français d'Afrique Noire* XXVI, série A, pp. 989–1022.

Crépy, M. (2016). "Les Paysages du vent : géohistoire et géoarchéologie de la dépression de Kharga (désert Libyque, Égypte) du cinquième siècle avant notre ère à nos jours - 2 500 ans d'interactions entre dynamiques éoliennes et activités humaines dans un milieu hyperaride." Thesis. Université Lumière Lyon 2.

Finkel, H. (1959). "The barchans of Southern Peru." *Journal of Geology* 67 (7), pp. 614–657.

Fryberger, S., A. Al Sari, T. Clisham, S. Rizvi, and K. Al Hinai (1984). "Wind sedimentation in the Jafurah sand sea, Saudi Arabia." *Sedimentology* 31, pp. 413–431.

Gallois, J.-P. (1965). "La lutte contre l'ensablement." *Revue d'Information de l'Organisme Saharien* 17 (September-October 1965), p. 14.

Glennie, K. (1970). *Desert Sedimentary Environments*. Amsterdam: Elsevier. 222 pp.

Hashtenrath, S. (1987). "The barchan sand dune of South Peru revisited." *Zeitschrift für Geomorphologie*. N.F. (31), pp. 167–178.

Inglethorpe, S. (1993). *Industrial Minerals Laboratory Manual: Diatomite*. BGS Technical Report WG/92/39. Keyworth, Nottingham, British Geological Survey. 37 pp.

Inman, D. (1952). "Measures for describing the size distribution of sediments." *Journal of Sedimentary Petrology* 22, pp. 125–145.

Kok, J., E. Parteli, T. Michaels, and D. Bou Karam (2012). "The physics of wind-blown sand and dust." *Reports on Progress in Physics* 75, p. 106901.

Long, J. and R. Sharp (1964). "Barchan-dune movement in the Imperial Valley, California." *Bulletin of the Geological Society of America* 75, pp. 149–156.

Mainguet, M. (1970). "Un étonnant paysage : les cannelures gréseuses du Bembéché (N. du Tchad). Essai d'explication géomorphologique." *Annales de Géographie* 79, pp. 58–66.

— (1972). *Le Modelé des grès : problèmes généraux*. Paris: Institut Géographique National. 657 pp.

Mainguet, M. and Y. Callot (1978). *L'Erg de Fachi-Bilma (Tchad - Niger) - contribution à la connaissance de la dynamique des ergs et des dunes des zones arides chaudes*. Paris: Mémoires et Documents du C.N.R.S. 18. 184 pp.

Marticorena, B., G. Bergametti, B. Aumont, Y. Callot, C. N'Doume, and M. Legrand (1995). "Modeling the atmospheric dust cycle: 2. Simulation of the desert dust sources over Sahara and Sahel." *Journal of Geophysical Research* 102, pp. 4387–4404.

Norris, R. (1966). "Barchan dunes of Imperial Valley, California." *Journal of Geology* 74, pp. 292–306.

Ouerchefani, D. (2012). "Caractérisation et suivi des états de surfaces éolisés en Tunisie pré-saharienne : approches stationnelle et spatiale." Tunis El Manar Univrsity and Université Lumière, Lyon 2. 354 pp.

Oulerhi, T. (1992). *Étude géodynamique des migrations de sable éolien dans la région de Laayoune (Nord du Sahara marocain)*. Mémoires des Sciences de la Terre 92-12. Université Pierre et Marie Curie (Paris 6). 223 pp.

Ping, L., C. Narteau, Z. Dong, Z. Zhang, and S. Courrech du Pont (2014). "Emergence of oblique dunes in a landscape-scale experiment." *NG* 7, pp. 99–103.

Rognon, P. (1994). *Biographie d'un désert*. Paris: L'Harmattan. 350 pp.

Roux, M. (1996). *Le désert de sable : le Sahara dans l'imaginaire des Français (1900-1994)*. Paris: L'Harmattan. 206 pp.

Sarnthein, M. and E. Walger (1974). “Die äolische Sandstrom aus der W. Saharazur Atlantikkuste.” *Geologische Rundschau* 63, pp. 1065–1087.

Sebe, K. et al. (2011). “Wind erosion under cold climate: A Pleistocene periglacial mega-yardang system in Central Europe (Western Pannonian Basin, Hungary).” *GEOM* 134, pp. 470–482.

Silvestro, S., L. Fenton, D. Vaz, N. Bridges, and G. Ori (2010). “Ripple migration and dune activity on Mars: Evidence for dynamic wind processes.” *Geophysical Research Letters* 37 (20), p. L20203.

Smith, R. (1970). “Barchan dunes in a seasonally reversing wind regime, Southeastern Imperial County, California.” M.S. thesis. University of Arizona, Tucson.

Thomas, D. (1992). “Desert dune activity : concepts and significance”. *Journal of Arid Environments* 22, p. 31-38.

Thomas, D. (1997). *Arid Zone Geomorphology*. London: John Wiley & Sons Ltd. 713 pp.

Tsoar, H. (1974). “Desert dune morphology and dynamics, Al Arish.” *Zeitschrift für Geomorphologie*. Supplement Band (20), pp. 41–61.

— (1982). “Internal structure and surface geometry of longitudinal seif dune”. *Journal of Sedimentary Petrology* 52, p. 823-831.

Twidale, C. R. (1980). “The Simpson Desert, central Australia.” *South African Geographical Journal* 65 (1), pp. 3–17.

Venard, C. (2012). “La morphodynamique éolienne en régions sèches - des échelles spatiales aux outils.” Université Lumière, Lyon 2. 328 pp.

Vermeesch, P. and S. Leprince (2012). “A 45-date time series of dune mobility indicating constant windiness over the central Sahara.” *Geophysical Research Letters* 39 (4). doi: 10.1029/2012GL052592.

Warren, A. (1972). “Observations on dunes and bi-modal sands in the Ténéré Desert.” *Sedimentology* 19, pp. 37–44.

Wells, S., J. Dohrendwend, L. McFadden, B. Turrin, and K. Mahrer (1985). “Late Cenozoic landscape evolution of lava flow surfaces of the Cima Volcanic Field, Mojave Desert, California.” *Bulletin of the Geological Society of America* 90, pp. 1519–1529.

Wells, S. and J. Zimbelman (1997). “Extraterrestrial arid surface processes.” *Arid Zone Geomorphology*. Ed. by D.S.G. Thomas. London: John Wiley & Sons Ltd., pp. 659–690.

Wilson, I. (1971). “Sand waves – Universal discontinuities in bedforms produced by wind.” *Geographical Journal* 137, pp. 180–189.

— (1972). “Aeolian bedforms - Their development and origins.” *Sedimentology* 19 (3-4), pp. 173–210.

— (1973). “Ergs.” *Sedimentary Geology* 10, pp. 77–106.

Zhang, D., C. Narteau, O. Rozier, and S. Courrech du Pont (2012). “Morphology and dynamics of star dunes from numerical modelling.” *Nature Geoscience* 5, pp. 463–467. doi: 10.1038/ngeo1503.

# 2

# Humans and Winds in Deserts

The relations between humans and the wind are sometimes difficult and complicated, as we saw in the Introduction. It must be recalled that for societies that live in these arid regions, the chief cause behind the setbacks brought about by the wind is of the wind's relative independence vis-à-vis the topography. The movement of water, governed by gravity, follows the lowest points on the land. The wind, however, is changeable, ubiquitous and varies in different places in the same region, depending on the relief. Without wind measurements, followed by calculating, or observing its outcome in the field, there may be significant errors arising from erroneous interpretations of aeolian formations themselves.

The impacts of an effective wind, which are well understood by sedentary farms in deserts and along their margins, are so poorly known that they generate any number of countertruths or total errors. In order to demonstrate the complexity of the interactions, we will look at one example of traditional management, a successful method used in Mauritania, using very few technological advances, and two examples of failed management: one from one of the world's richest countries, the United States, and the other in a resource-rich country in the Maghreb, Algeria.

---

Chapter written by Yann CALLOT. All photographs within this chapter have been provided by the author, unless indicated otherwise.

## 2.1. Traditional societies and the wind

For millennia, traditional societies have accommodated the winds and sandy invasions, accumulating knowledge based on the observation of their natural environment, something that is, alas!, disappearing today. In the Laayoune region, Oulerhi (1992) observes:

> The lifestyle changes in the past 60 years have changed the knowledge of barchans among the inhabitants of the region. We learnt from a resident of Foum el Oued that the barchans were earlier used as granaries or to hide goods away from the *razzias* [raiders]. People would bury their goods at the foot of a chosen barchan based on its size and on the season, depending on whether they were making a "long-term deposit or a short-term deposit".

This anecdote is a reminder that we must not neglect non-geodynamic accounts and arguments when studying these phenomena.

Certain ancient techniques are so effective – and cost-effective! – at protecting against the sands that they are still very much in use today. The simplest are small hedges surrounding plots of land to protect them. This technique has been used for many centuries as evidenced by their presence on the sanded-over medieval site of Sedrata, near Ouargla (Algeria). These hedges are found in almost all deserts, whether as a traditional, artisanal form of protection or in larger installations, criss-crossing entire dunes (Figure 2.1).

One of the most striking techniques is the use of a protective, rectilinear plant barrier, most often made up of palms, though other plants are used in regions with no palm trees. In Maghrebi Arabic, a dune covered in these barriers is called an *afreg* (Figure 2.1), and this technique has been studied by Bisson (2003) in the Badriane village (Gourara, Algeria):

> Thus, to protect their gardens and their villages, the residents have always carefully erected a fence of dry palm fronds, which blocks most of the sand in transit [...] over the years, this sand accumulates and the dune grows in size, its base grows broader (making it necessary to plan for some space between the fence

and the construction or infrastructure that is to be protected) and it stabilizes, as long as the fence, now located on the summit of the dune, is carefully maintained.

**Figure 2.1.** *Fixation of dunes on the left bank of the Niger, east of Timbuktu (photo Éric Delaître). For a color version of this figure, see www.iste.co.uk/bourrie/deserts3.zip*

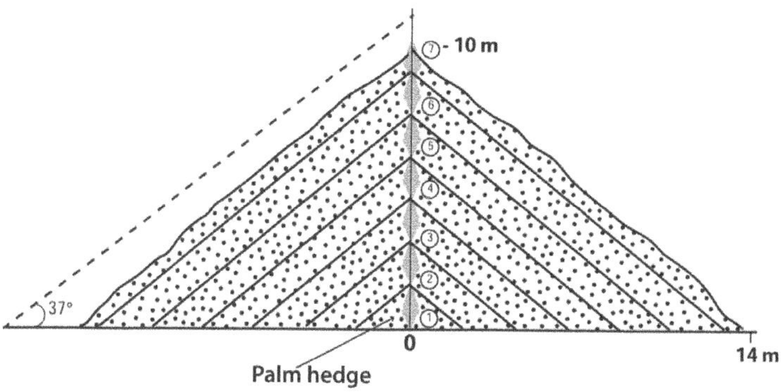


**Figure 2.2.** *Cross-section of an afreg*

Although an afreg is not expensive, as it is constructed from local vegetation, it does require intensive labor, earlier furnished by the *Touiza*, a voluntary workforce formed by residents of the village. Set up in an appropriate location, the afreg is effective as it is an evolving structure. Figure 2.2 shows that it is initially quickly filled by a small quantity of sand. However, as it continues to grow higher, it becomes more and more effective as the volume of sand that is blocked increases as a cubic function, while the area of the palm-frond screen is a square function. When each hedge is 2 m high, the first row of palm fronds can block about 5 m$^3$ of sand over a width of 1 m, the second row can block about 15 m$^3$ more, the third row can block 30 m$^3$ more, the fifth row can block 55 m$^3$ and so on. Since the highest of these fences have a height of 30 m, they can thus block about 1000 m$^3$ per linear meter, or 1 million m$^3$ per kilometer. When well maintained, they can protect entire oases, as is the case in Akabli (Figure 2.3), to the SE of Reggane (Algeria) where a series of afregs are arranged facing the prevailing NE wind.

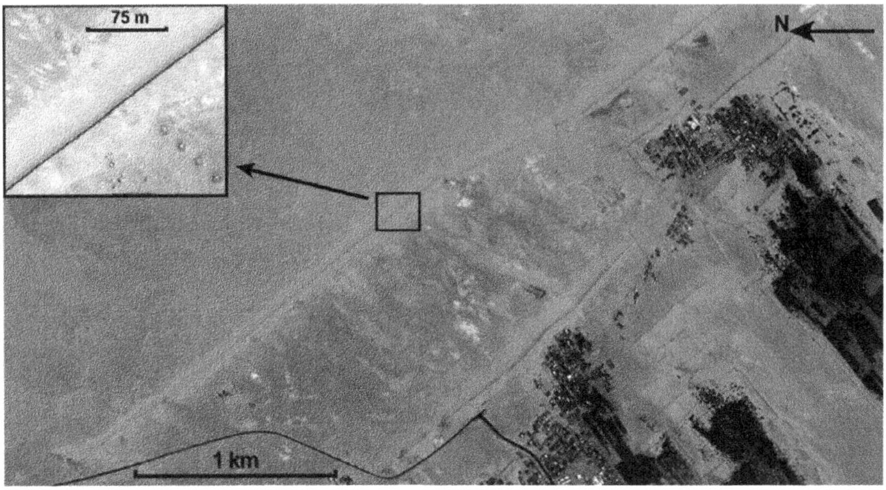


**Figure 2.3.** *Afregs protecting the Akabli oasis (Algeria). The first dune is over* 3 km *long. As it grows it can be seen that it attains a width of* 80 *to* 90 m *for a height of* 30 m. *The dune also covers a series of foggaras (wells), which is a sign that this is a recent dune. The regularity of the black lines of palms shows that this fence is well maintained. (Google Earth Image)*

The largest of these is 4.4 km long, and groups of these structures extend over 10 km. However, it becomes increasingly difficult to maintain these dunes when they reach this size and when they are abandoned, or simply neglected, breaches open up, allowing sand through (Figure 2.4).

The dying away of the Touiza, especially its absence in recent installations, makes the upkeep of the fences a costly affair. However, roads that are threatened by advancing dunes are still often protected (for example in southern Tunisia or the Adrar region in Algeria) by afregs.

**Figure 2.4.** *Afregs in the Douz region (southern Tunisia). The dune is about 15 m high and has been abandoned. It is disintegrating, revealing the lower palm-hedges, and its edges have become irregular. For a color version of this figure, see www.iste.co.uk/bourrie/deserts3.zip*

## 2.2. The battle against sand encroachment, or the Barrel of the Danaids: the Zouerate railroads (Mauritania)

The railroad track stretching across 700 km, connecting Nouadhibou with the iron mines in the Zouerate regions, NW of Mauritania, is an excellent example of the problems that arise at the interface between human society and a hostile environment (Figure 2.5).

Given the important role it plays in the local and national economy (Dejouhanet 2006), it is essential to overcome the constraints imposed by the desert environment. Over 400 km, the railroad is oriented W-E along the border, while the wind migration, going from NNE-SSW close to the sea and NE-SE to the east, cuts across the railroad at an angle that varies between 80° and 45°. These migrations are especially large, as we have seen when studying barchans. The movement of trains is made more difficult by certain characteristics of railroads:

– sand infiltration into the ballast causes it to lose the "elasticity" that helps diminish a train's vibrations. For a train transporting over 20,000 tons of minerals in addition to the weight of the wagons and engines, a derailment would be catastrophic both economically as well as in human terms, as people travel in the carriages that carry iron ore, whether they are empty or loaded;

– the sand is a loose and terribly abrasive rock that wears out the rails and blocks the switches. Being completely incompressible, a layer of just 5 cm on the tracks is enough to derail a train;

– if unchecked, the migration of the barchans brings in considerable quantities of sand onto the railroad, which can completely block traffic.

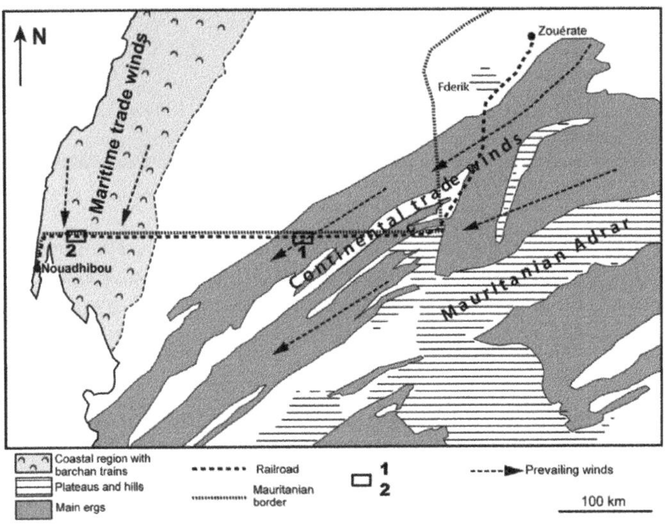


**Figure 2.5.** *Location of the Nouadhibou–Zouerate railroad. The zones marked **1** and **2** refer, respectively, to Figures 2.6 and 2.7(b)*

The railroad maintenance services have very few means of battling this movement, as the wind is so powerful in this region that it can move between 150 and 200 tons of sand per year and per linear meter onto the western end of the railroad, where the maritime trade wind blows. The quantity moved is probably lower in the eastern part, where the continental trade wind is less violent. Given this mass of material, the mining company engineers have used various techniques to keep the trains running.

At one specific point along the railroad there is a hard rock-ledge that is slightly oblique in the direction of the prevailing wind, that channelizes the sand, which continues to accumulate here. Teams of workers operate in shifts, night and day, to clear the rails by shoveling the sand out to the leeward side of the rails, from where it is carried away (Figure 2.6).

**Figure 2.6.** *On the Nouadhibou–Zouerate railroad, workers constantly clear one part of the tracks where the sand always encroaches. For a color version of this figure, see www.iste.co.uk/bourrie/deserts3.zip*

In this case, the wind itself is used to help in clearing the tracks as the sand shoveled out, just to the side of the tracks, is naturally carried away. On the western part of the tracks, the SNIM (the mining company) engineers used the same natural energy to help remedy the problem of shifting hundreds of thousands of tons. Faced with very large barchans approaching the railroad, they searched for and found a way to not only "block" the barchans but also destroy them, with minimal effort. Their highly pragmatic approach was based on a series of trials until they found the best cost/benefit ratio. In order to block the dunes, they regularly lay down heaps or bands of "pit run gravel" on the dune (Figure 2.7).

As we have seen, the wind cannot move gravel or pebbles and thus hollows out the area around the heaps, which form perched structures. The dune "melts" in the same place and the sand, carried away by the wind, passes over the railroad without blocking it. This technique makes use of the differences in scale, knowing that the wind's behavior differs from grains that are millimetric to centimetric gravel, in order to destroy structures on a hectametric scale without expending any effort. It is a remarkably smart strategy and must be lauded.

(a) Overall view of the dune. The person is on the railroad, which can be seen on the left. Through erosion, a mass of soil equivalent to the height of the protuberances has already been carried away.

(b) The protuberances on the summit of the dune. Through erosion, the soil between the protuberances is carried away and the original heap "flows down" until it forms a completely covered cone.

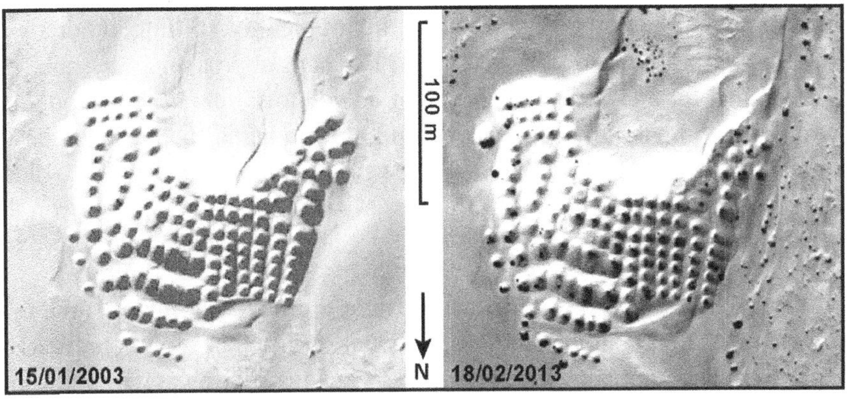


(c) Comparison of two Google Earth images, taken 10 years apart, of a barchan where the "heaping technique" has been used. The dune has not moved, but no longer releases soil as it is practically covered by the debris from the heaps.

**Figure 2.7.** *Techniques for destroying a barchan that threatens the Nouadhibou–Zouerate railroad to the west of Boulanouar. For a color version of this figure, see www.iste.co.uk/bourrie/deserts3.zip*

### 2.3. The Dust Bowl, the first ecological disaster of the 20th century in a developed country

Certain aeolian formations can offer remarkable agricultural potential while also presenting grave risks. This is the case of loess, already mentioned in the second part of the study on wind, which is chiefly of periglacial origin in Eurasia and North America (Antoine and Lautridou 2008). Loess makes up superficial formations that cover vast areas in the middle latitudes of the northern hemisphere. Formed from the deflation of the periglacial areas around ice-sheets, they are carried on the wind and, consequently, are very susceptible to being carried away again by the wind once deposited.

The point of convergence between the erosive power of aeolian actions in an arid climate and technological progress occurred in a logical manner in the United States. A large part of the plains in the center of this rich country is covered by loess (Figure 2.8).

The loess is thick in the northern and eastern part of the plains, and thinner in the western part, which is often semi-arid. In the 1930s, agriculture had already become quite mechanized. In the dry, sometimes semi-arid, regions in the large plains in the Midwest, the additional labor that this mechanization provided made it possible to cultivate vast areas using agricultural practices that periodically destroyed the deflation pavements (seen earlier), exposing bare soil. The Midwest was often covered in loess soils. The system, which had been in equilibrium during the humid years, was thrown into imbalance by a decade of droughts starting in 1930. After a few smaller incidents, the extent of the catastrophe was driven home on Black Sunday, April 14, 1935 when an enormous "dust front", called a black blizzard, descended, turning the day into the darkest night (Figure 2.9; NOAA 1935).

The term *Dust Bowl* was coined the next day. The wind carried the majority of the fine particles over millions of hectares of land that was bared and exposed by agricultural activity, sometimes in the form of enormous clouds of dust, while other areas saw considerable accumulations. Hundreds of thousands of people, ruined by the loss of their land, their crops buried under dust, were forced to leave the region. Many of them emigrated to California (Figure 2.10).

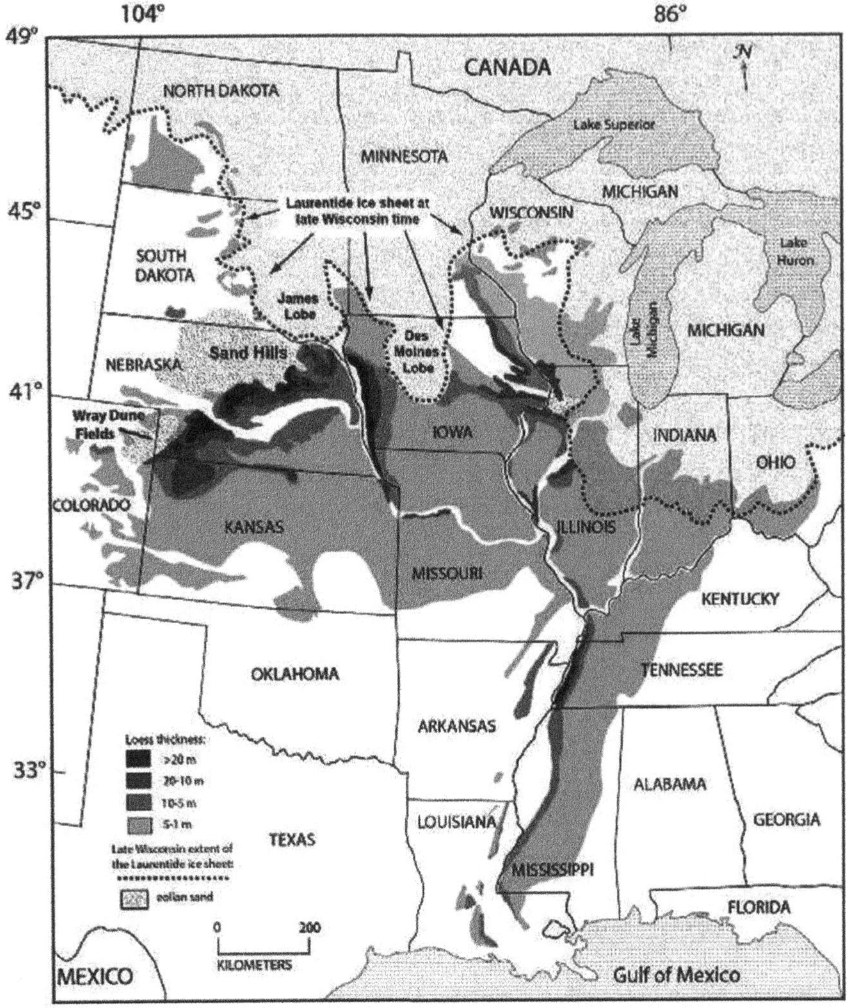


**Figure 2.8.** *Extent and thickness of loess in the United States with respect to the area covered by the Laurentide ice-sheet (as per Bettis III 2012)*

Their fate was the subject of John Steinbeck's famous novel, *The Grapes of Wrath*, and the film of the same name by John Ford. Authors state that the phenomenon extended over a wide area whose limits are not clear; it extended from north to south for over 2,000 km (Figure 2.10), and the processes of ablation and accumulation alternated so closely that there was no way to identify which areas experienced chiefly ablation and those where accumulation predominated (Figures 2.11 and 2.12), if this can even be measured.

(a) The city before the dust storm

(b) The city 15 minutes after the dust storm started

**Figure 2.9.** *Two "identical" photographs of Garden City (Kansas) on Black Sunday (April 14, 1935). Only the street lamps which were on and visible in both photos (a) show that the two pictures are of the same site (NOAA 1935)*

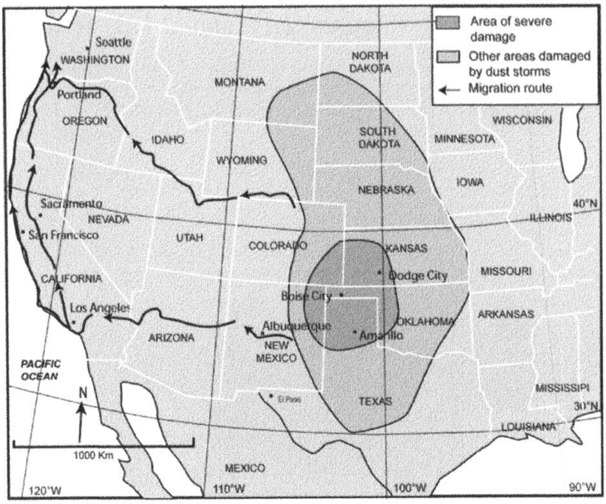


**Figure 2.10.** *Extension of the Dust Bowl and the migrations associated with the phenomenon (source: mrsreesehistory.weebly.com). Dark gray: the area with severe damage; light gray: other damaged zones; arrows: migration routes*

**Figure 2.11.** *A farmer watching a dust storm in Cimarron county, Oklahoma (photo by Arthur Rodstein/Library of Congress)*

One of the reasons for the magnitude of this early 20th century environmental disaster is the remarkable fertility of the loess, which motivated people to settle down in climatic conditions that bordered on the extreme in an era where there was no knowledge of environmental issues and where the harmful effects of overcultivation, especially the lack of protection of the soil, were unknown. The irony was that once the dust storms died away, the lands were covered in fresh deposits which, once fixed, made them more fertile. Analyzing the area on Google Earth, for example, shows that today the plains are rich and fertile, especially in the northern part of the region.

**Figure 2.12.** *A farm buried under aeolian deposits in Dallas (South Dakota) in 1936 (source: United States Department of Agriculture)*

## 2.4. The western Algerian Green Barrier, or an error in management

### 2.4.1. *A "barrier" in an unfavorable physical environment*

Algeria is a country with a single climatic zonation: the droughts increase overall from north to south. The phenomenon is exacerbated, but sometimes locally reversed, by relief that is parallel to the climatic lines. In western Algeria or southern Oran, which will be studied more closely as the contrasts are most pronounced here, we can see (Figures 2.13 and 2.14):

– a thin coastal band with low humidity, where precipitation is around 300 mm/yr in Oran;

– the Tellian Atlas, the most humid: 400 mm/yr to 500 mm/yr, locally higher where the orographic effect is greatest;

– the High Plains – which were called the High Plateaus for many years, due to an error on the part of French geographers who were part of the colonists – are quite dry, especially in their central region (around 200 mm/yr);

– the Saharan Atlas (which is not really in the Sahara... those geographers strike again!) where the precipitation again increases from 250 mm/yr to 300 mm/yr through the orographic effect on the septentrional slopes;

– the large hamadas in the Sahara proper. The difference in altitude between the two piedmonts of the Atlas provoke a foehn effect, which quickly brings about arid conditions in the southern piedmont of the Atlas: the precipitation here is barely more than 100 mm/yr at the foot of the chain and decreases rapidly toward the south.

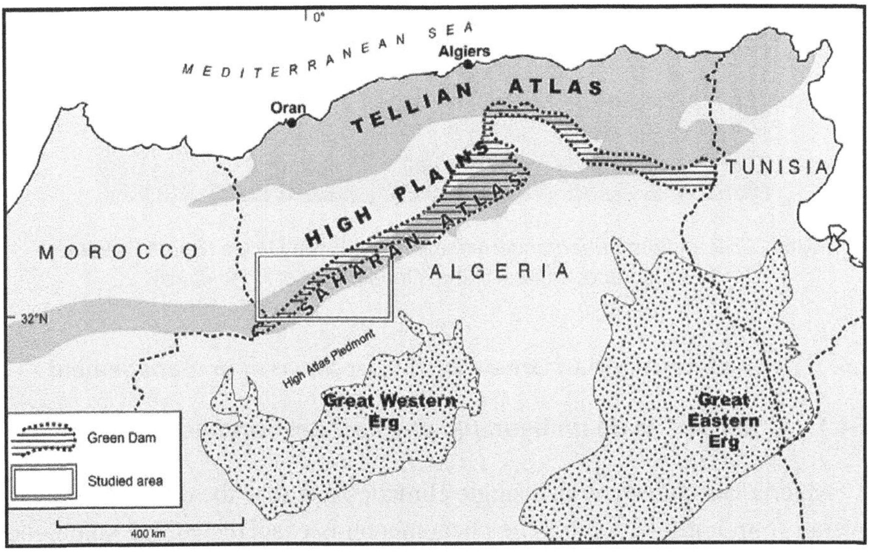


**Figure 2.13.** *Septentrional Algeria. The Green Barrier was chiefly established at the point of contact between the Saharan Atlas and the High Plains*

This territory sees little agricultural activity in the High Plains–Saharan Atlas region, and none at all on the meridional piedmont. However, livestock rearing, especially sheep farming, is an important activity in the region, except in the Sahara.

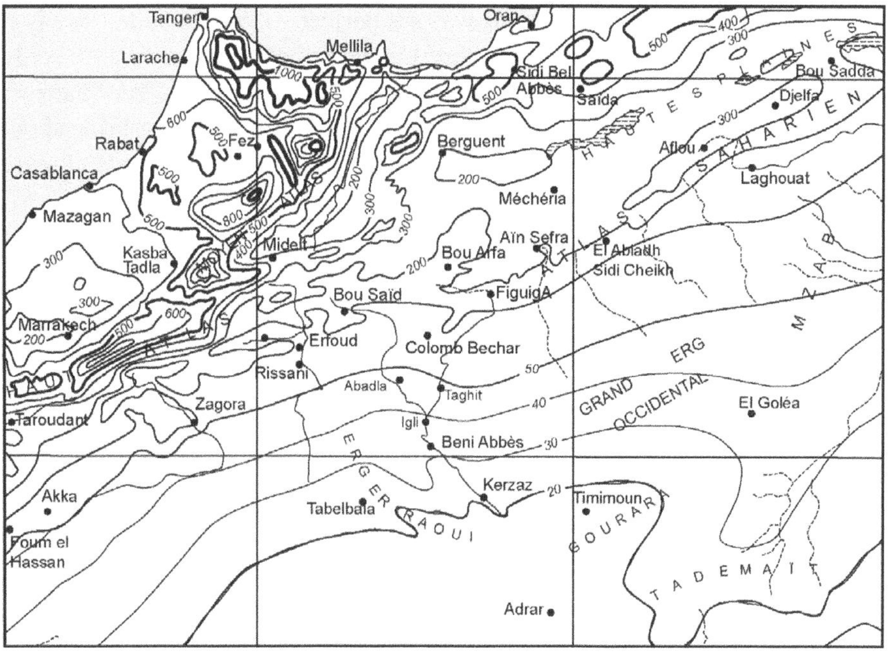


**Figure 2.14.** *Precipitations in western Algeria and Morocco, as per Dubief 1959–1963. Present-day values are generally lower*

### 2.4.2. *Difficult implementation*

Environmental degradation has been high for a long time throughout this region. For decades now, the High Plains have seen an increase in the cover of active sand: small nabkhas appear in increasing numbers toward the north. There is notable desertification with a decrease in green cover, especially on the Alfa steppe, which covered a large part of the plains.

In 1969–1970, under the initiative of the Algerian President Houari Boumediene, the massive *Green Barrier* project was launched. It was to extend, symbolically, from the Moroccan border to the Tunisian border, a distance of about a thousand kilometers. It was meant to prevent the northwards progression of sand and desertification. It was set up in the west on the septentrional piedmont of the Saharan Atlas in sectors where the precipitation is 150–200 mm/yr. In the eastern part of Algeria, the more complex relief led to a less simple distribution, which would not be taken up.

The location of the western part was not perfect – it would be fair to say it was the better option – but the choice was logical and pragmatic: it was the only area where an increase in plant cover could successfully happen. The "barrier" was often created through plantations of trees on plots of lands measuring about 500 m in length and 100 m in width, aligned along the major roads (Figure 2.15).

Executed under the army's direction, the operation was a large one. The labor was made up of many conscripts and members of voluntary organizations. However, the western part of the operation was an overall failure, for many reasons:

– the almost universal use of a single species of tree, the Aleppo pine, chosen because it was particularly well adapted to the harsh climate in the Atlas piedmont, which was too dry for other trees, very cold in winter and very hot in summer. This made the Barrier vulnerable to attacks by pine parasites: an invasion of pine processionary caterpillars ravaged certain sectors that was lacking in any other kind of plant species that could have mitigated the attack;

– the trees were planted on long strips aligned with roads, chosen for logistical reasons. There was, thus, no selection of areas based on which would be the most favorable for plantation. Further, the climate is the most extreme the trees can tolerate and if were any more arid the trees would die. This made them particularly sensitive to local constraints. As soon as negative factors, even the smallest ones, appeared, the plantations would die out. For example, all the pines died along certain interfluves due to a lack of any additional water and in the near-absence of soils. In other areas, below a slope that provided additional water from upstream, for example, the plantations succeeded. Certain talwegs can clearly be seen on high-definition satellite images, through a central part that is lacking trees, torn through by floods and a line of trees that are much better developed than elsewhere because of the water brought in by the wadi, forming a "micro riparian forest" (Figure 2.15(b)).

In this kind of an environment, aeolian sands can play an important role: if they are not overabundant, blocked by the roughness of the trees, the sands can accumulate at the foot of the trees and form small nabkhas, often joined up together to form sandy nappages. The sand then traps rainwater. Its water retention capacity and its ability to block runoff is much more effective than that of the soils on the southern Oran steppes (Figures 2.15 and 2.16).

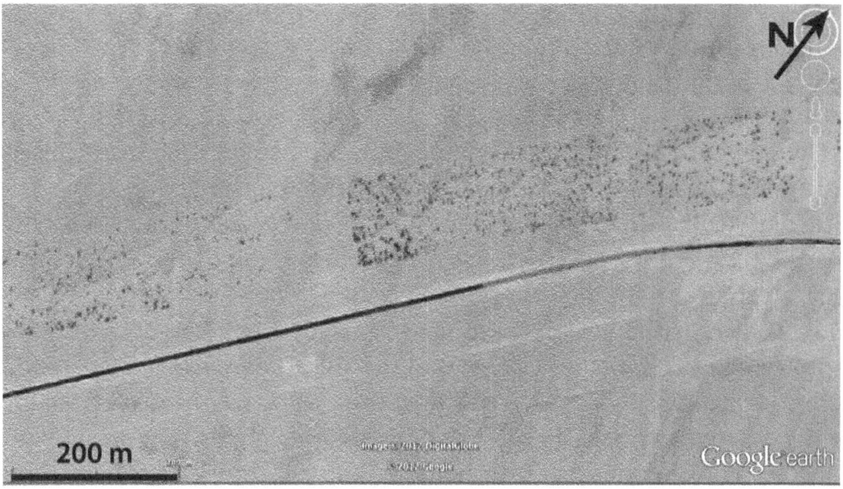


(a) Position of the plots of land (200 m scale). These follow the road, lying to its north. The plot on the right is well preserved and the plot on the left has almost disappeared in parts (image taken on July 24, 2003).

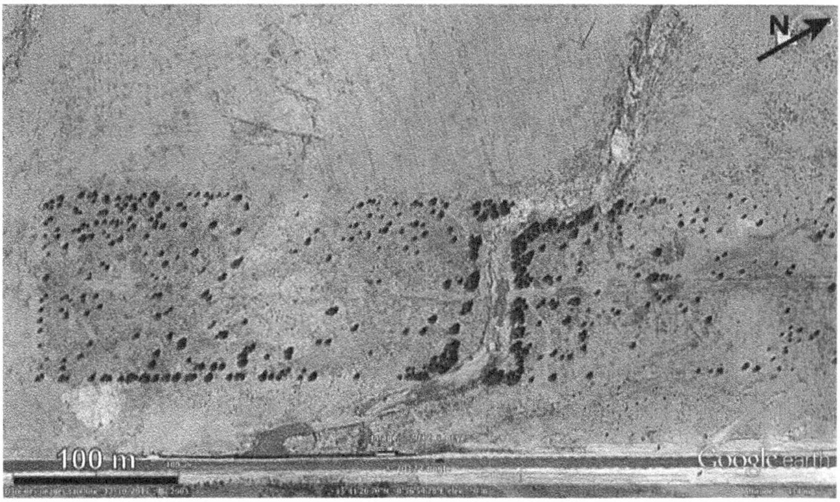


(b) Detailed view of the contrast between different parts of a plot of land located 4 km to the east of the previous one (100 m scale): a small wadi goes through the lot and its bed is completely lacking in trees, carried away by the floods. However, on its banks you can see a burgeoning riparian forest that contains the most developed of the trees on that land. Trees have totally disappeared on certain clear spots (probably resulting from crusts forming) (image taken on October 12, 2011).

**Figure 2.15.** *The lands of the Green Barrier, near the city of El Bayadh, in a sector where there has been good conservation of the plots (Google Earth Images from October 2003). For a color version of this figure, see www.iste.co.uk/bourrie/deserts3.zip*

Along the borders of talwegs, these inflows sometimes add to the water brought in, creating favorable conditions as can be seen in Figure 2.16(b) where there is equilibrium between inflow and growth.

However, if too much sand is brought in, it smothers the trees, burying and killing them. As sand deposits are often due to the growing trees disrupting aeolian flux, the predominance of one action over the other has long been a precarious position, even if we may think that when the plants reach a certain height they are less likely to be "drowned" by encroaching sands. If there is a large amount of sand, it can destroy trees, even those that are full-grown, as sometimes happens in the olive groves in southern Tunisia. Further, a plantation that survived for several years may be quickly snuffed out if a drought increases the amount of sand brought in, when tree growth is stunted by climatic stress.

A simple study using Google Earth shows that the Green Barrier was a failure in the southern Oran, even if certain sections thrived, especially around El Bayadh. We cannot estimate the magnitude of the failure, as there is a lack of numerical data on the percentage of area that saw success, that saw partial success and areas where almost nothing is left anymore apart from cairns on the four corns of the plot, broken up by the plantation ditches. Another event further reduced the project's chances of success: in the years from 1970–1986, the region suffered a severe drought, the most severe and the longest known (Bisson and Callot 1990). Would the project have succeeded if this drought had not occurred? It is not certain in such a difficult environment. The importance of the "precipitation factor" was confirmed by another factor: this region is the driest of all regions where the Green Barrier project was implement (Figures 2.13 and 2.14). Further to the northeast, toward Djelfa, there is slightly higher precipitation over 300 mm/yr. This is why the implementation of the Barrier project was not subject to the same constraints in this region and larger wooded areas exist here today, either through the conservation of natural forests or through reforestation. These areas are particularly large between 2°30′ and 4°30′E.

(a) Google Earth Image (27 July 2007) of a sector where certain areas are completely covered in trees and others are completely bare of them.

(b) Sector with high sand encroachment, but where the trees have grown well.

(c) A transition sector between an area where a few trees have been maintained and an area where there is nothing left of the plantation. Photo taken on the ground in March 2007.

**Figure 2.16.** *The Green barrier in the Naama region, Algeria (High Plains in the southern Oran). For a color version of this figure, see www.iste.co.uk/bourrie/deserts3.zip*

All this confirms that the southern Oran was a region where the most extreme conditions of survival for pines across the Barrier led, in most causes, to the trees dying out.

### 2.4.3. *A basic error in understanding aeolian dynamics*

While the creation of the Green Barrier met with many difficulties, the theoretical basis of its creation itself included a basic mistake, chiefly resulting from a poor understanding of aeolian dynamics. It has already been seen that the northward shift of the border of regions with active sands and, in general, the increase in the area of surfaces where sand has encroached throughout the region were two reasons for the construction of this barrier. Based on these facts, the contexts of that era favored the decision to construct the Barrier and its continuation. The factors considered included:

– in the Sahel, extensive droughts in the years 1970 to 1980 drew people's attention, showing that there was a significant southward advance of active sands. In a desire for symmetry on either side of the Sahara (*Sahel* means "shore" in Arabic, here the shores of the desert) a "matching" phenomenon was seen in the advance of the sands in the High Plains;

– the Algerian State's desire, shortly after independence, to find a national cause that would bring together the population of a young state was a non-"nature" related factor that played into the decision. A barrier stretching from Morocco to Tunisia was an ideal project for such an initiative. For many years, official newspapers sang the praises of this barrier, a sturdy defense against the encroachment of Saharan sands into "habitable" Algeria;

– the presence of sandbanks from deposits from the south.

Things were different in reality, however. The "symmetry" with the Sahel was inaccurate. The very real encroachment of sand into this region was actually due to a combination of three factors:

– the abundant sandy substrate of the active dunes in the *Ogolian Erg*. This paelo-erg was formed during the time of the Sahara's greatest extension, around 18,000 years B.P., when it stretched 500 km further south of its current border (Rognon 1994). The desert sands covered the majority of the zone of contact between the driest part of the Sahel and the proper Sahara itself;

– the omnipresence of the harmattan in the Sahel. This continental African trade-wind blows from NE to ENE, thus from the interior of the desert toward its borders. This results in a migration of sand from the desert toward the periphery of the deserts, always at an angle, although this may be small;

– consecutive droughts that hit the Sahel from 1970 to 1990.

The wind regimes in the area of contact between the Sahara and septentrional Algeria were not well studied, perhaps because of the symmetry assumed between the northern and southern Sahara. Large sandbanks that turned towards the south on the slopes of the most meridional relief in the Atlas were (rightly) perceived as being formed by winds from the south. However, these formations are due to the sheltered position of theses slopes against the winds from the north. At the foot of these slopes, all aeolian forms indicate that the displaced sand end up NNW-SSE. A perfect example of this is the Arbouat Cluse (a valley running through a fold) north of El-Abiodh-Sidi-Cheikh (Figures 2.17 and 2.18).

In this vast valley, the Oued Menhar, fed by a higher level of precipitation over the Saharan Atlas, travels through the Atlas from north to south and then opens onto the piedmont; its slope decreases and it deposits a large quantity of sand collected from the disintegration of the sandstone in the chain. After each flood, several times a year, the wind picks up the sandy material deposited in the bed and destroys the deflation pavement formed after the previous flood. The material is carried away by the wind exclusively southward. It is so abundant that it passes over a WE crest parallel to the wadi and spreads itself out on the southern slope of the crest. After traveling east for 4 km, the Oued Menhar curves toward the south through the narrow passage formed by the Kheneg Menhar (Kheneg is the Arabic word for Cluse and means a water gap) and continues in an SW direction. The eastern limit of the sandy zones is a rectilinear path, which starts from the kheneg and carefully follows the NNW-SSE direction of the migration of the sands. Closer views show barchans south of the crest (Figures 2.19 and 2.20) and corrasion grooves, as was discussed when we studied corrasion (Figure 1.6), are present on the rocks of the crest, showing identical directions of migration.

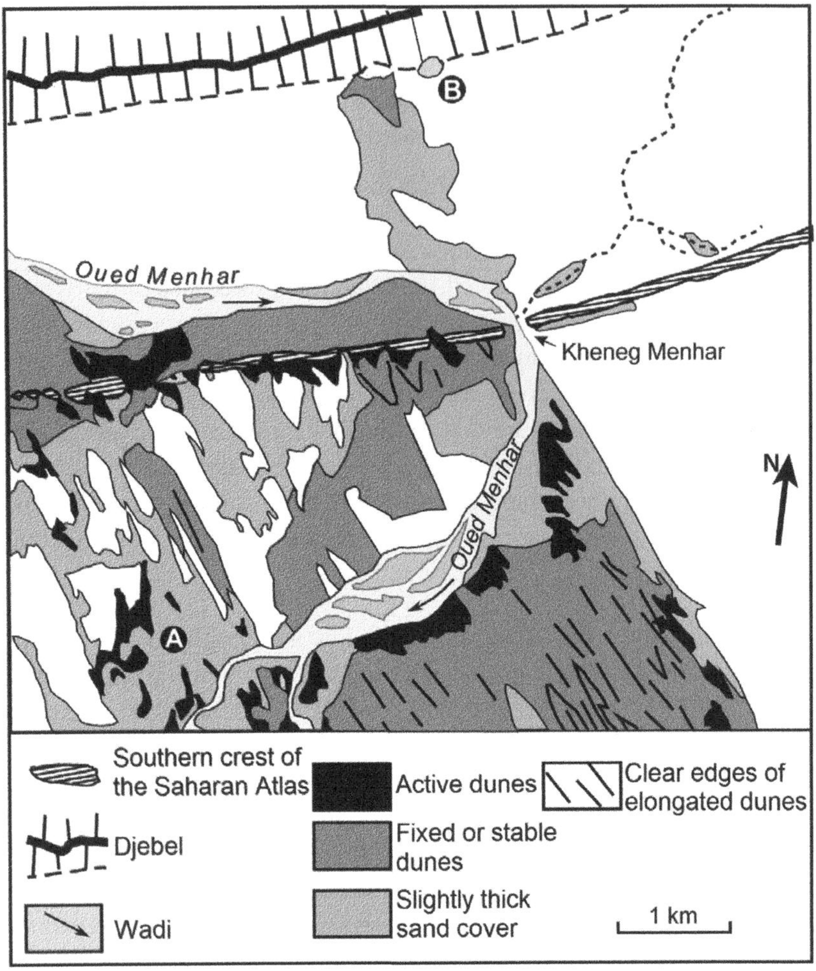


**Figure 2.17.** *Aeolian formations around the Kheneg Menhar (meridional dustfall from the Saharan Atlas, Algeria). Cartography based on aerial photographs, Mission ALG 1973 038/40, no. 246 0 248*

North of the Kheneg Menhar, however, a sandbank (marked B in Figure 2.17) can be seen from a great distance away because of its elevated position on the very steep southern slope of the Djebel dominating the Kheneg. This can in no way result from winds from the north.

The location of this sandbank can only be explained through a seasonal study of the potential sand migration roses from the station close to Aïn Séfra

(Figure 2.21). These indicate a variation in the direction of migration, expressed by the roses: in spring, autumn and winter, the sand migrations take place in an N-S to NW-SE direction, very dominant during winter and spring. These are so dominant that the annual migration rose has a clear NNW to SSE direction. However, in summer, it is less clear, and comes from the southern sector.

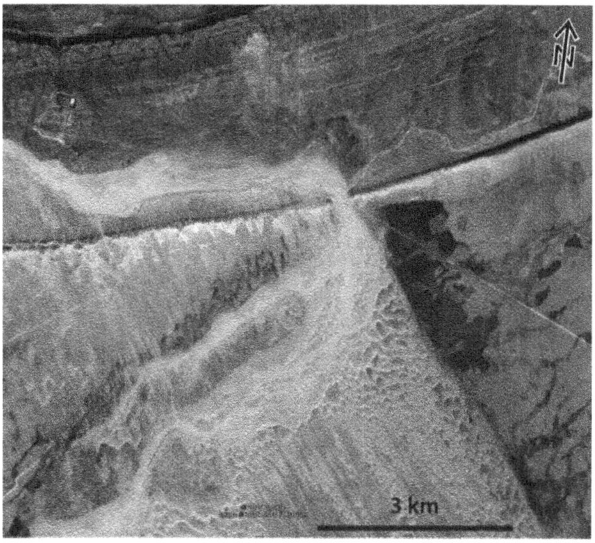


**Figure 2.18.** *Aeolian formations around Kheneg Menhar (the meridional skirt falling away from the Saharan Atlas, Algeria). (Google Earth Image from 2015). While dune formations may move, the areas with aeolian deposits have remained stable for 42 years. The frequent flows in the wadi (several times a year) deposit abundant sandy material coming from many upstream sandstone outcrops. This is picked up by the prevailing NNW winds and deposited only on the meridional bank wadi. Note the eastern, rectilinear border of the dune massif, corresponding exactly to the resultant winds. To the north, an isolated sandy outcrop forms a sandbank at the foot of the meridional slope of the Djebel. Sheltered from the winds from the north, this is fed by summer winds from the south. For a color version of this figure, see www.iste.co.uk/bourrie/deserts3.zip*

This particular feature makes it possible to explain the position of the sandbank: during this season, the Kheneg Menhar sands rise up toward the slope of the Djebel, where they are trapped, as this slope is totally protected from the prevailing winds from the north.

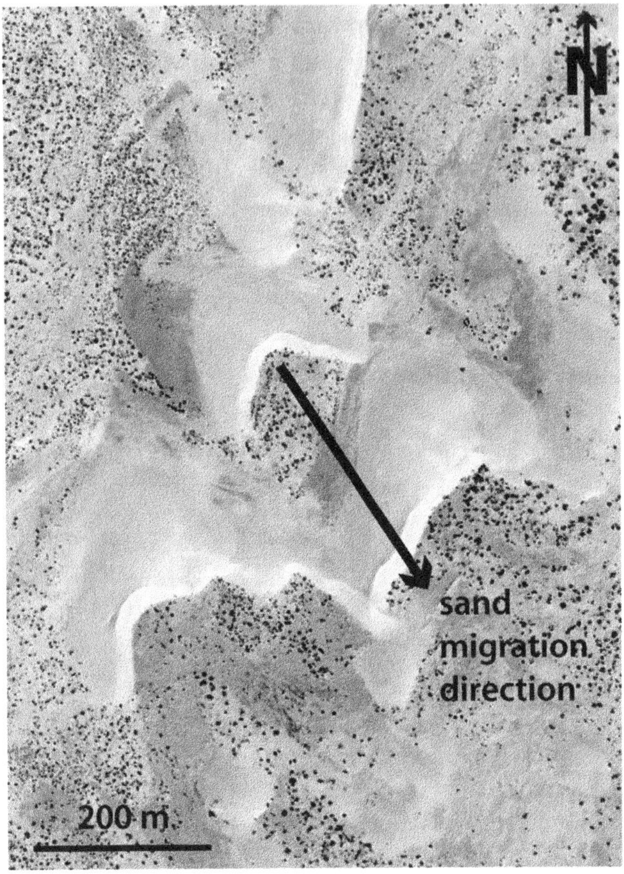


**Figure 2.19.** *Google Earth image of the coalescent barchans, SW of the Djebel Menhar. The barchans are in the direction of migration toward the SE, parallel to the eastern border of the dunefield visible in the previous image*

Its position is due to the conjunction of three very favorable factors:

– the presence of the Kheneg Menhar, which facilitates the passage of meridian winds by interrupting the protection offered by the meridional crest;

– the NNW-SSE orientation of the wadi in the Kheneg for over 1 km, which gives the wind a long area with a sand supply;

– the total protection of the southern slope of the Djebel against the septentrional winds.

**Figure □.□□.** *Directions of the migration of sands and the barchans of K heneg Menhar (Algeria). Barchans on the south of the crest of K heneg Menhar, NE of El-Abiodh-Sidi-Cheikh (Algeria). The view is of the SW. On the horizon, to the right, are the peaks of the Saharan Atlas. In the foreground, two coalescent barchans have a migration direction that is to the SE. For a color version of this figure, see www.iste.co.uk/ bourrie/ deserts3.zip*

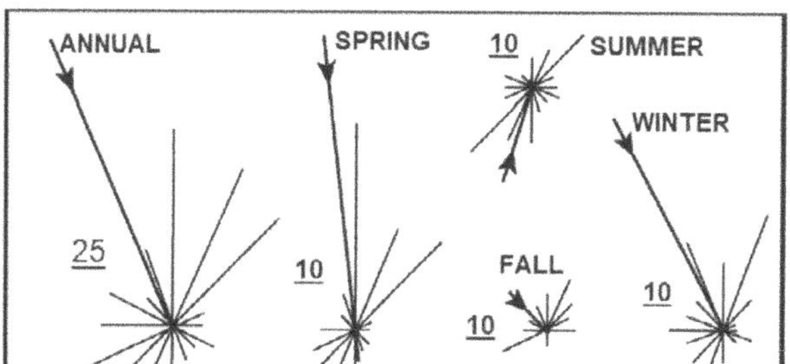


**Figure □.□□.** *Wind roses showing the potential migration of sands at the A□n Sé fra station for the period 1978– 1982. The thick lines with an arrow indicate the result of the migration and, therefore, the direction of displacement of sands and dunes. The annual rose is on a different scale from that of the seasonal roses*

The location of the sandbank, which is due to a unique situation, cannot be understood except by considering factors on a local, kilometric scale. There are sections that are so well protected that the prevailing wind here is from the south, although it is in the northern sector on a regional level.

The case of the Kheneg Menhar is far from unique. Many sites in the Saharan Atlas present the same arrangement of dunes and sandbanks, proving that **no sand crosses over the barrier of the Saharan Atlas to encroach into septentrional Algeria**. The north-to-south migration has been verified as far as the eastern part of Algeria, since Ballais, Marre, and Rognon (1979) already demonstrated it west of Biskra.

Thus, the Green Barrier has no sand to block here. What remains to be explained is the very real phenomenon of the northward progression of the border of the active sands. This must, surely, indicate a more prosaic factor: overexploitation of the High Plains steppes, chiefly due to overgrazing. The deterioration of plant cover from this was such that the fine and sandy part of the soil was made mobile. The process, which began in the middle of the High Plains, advanced toward the north. There was, thus, an advance of the desertification front that picked up local material, and not a displacement of the Saharan sands toward the north. What remains to be explained is why a project of such magnitude and involving such large expenses was based on incorrect science. There are many parameters that could have played a role: the primacy of theory over practice and the engineers' consciousness of their superiority, as the urban elite from non-arid regions, over the livestock farmers and peasants from the Saharan Atlas. These "locals" did in fact have a perfect knowledge of the direction in which the sands migrated, but they were never asked. In 1981, an illiterate *fellah* (Maghrib farmer) from Sidi-Ahmed-Bel-Abbès, a remote village in the middle of the Saharan Atlas, told us with conviction that the encroachment of sand onto his fields originated in the northwest.

Finally, and above all, the construction of the Green Barrier very soon became a political act, a "national cause" meant to cement the unity of a newly independent state within borders arbitrarily established by its former colonial rulers. While some people became aware of the error, it became politically impossible to admit it. Toward the end of the 1980s, certain senior officials in the Algerian administration privately shared with us their conviction that the "official" displacement of the sand was wrong, but added that they could never put forward such a suggestion in their "work environment".

Certain lessons from the failures of the Green Barrier were taken into account in the new attempts to restart the project from 1990 onwards. Information on the current state of the project is quite hard to come by, but analyses of Google Earth images show that while certain parts have succeeded, in central and eastern Algeria, especially around Djelfa, to the west of El Bayadh, the reforested areas have been reduced to points. The Algerian Green Barrier, a project that was carried out over decades based on an erroneous postulate, is one of the most remarkable examples of the importance of studying the wind in deserts and the necessity for dialogue between science and politics. It complements the examples seen earlier to show that in relations between humans and the wind, it is often not a case of a lack of money or resources that scupper projects, but a lack of intellectual humility before a complex natural phenomenon where both a researcher's knowledge as well as that of a *fellah* must be deployed.

## 2.5. References

Antoine, P. and J.-P. Lautridou (2008). "Les formations loessiques en milieux périglaciaires." *LES FORMATIONS SUPERFICIELLES, Genèse — Typologie — Classification — Paysages et environnement — Ressources et risques.* Ed. by Y. Dewolf and G. Bourrié. Paris: Ellipses, pp. 322–331.

Ballais, J.-L., A. Marre, and P. Rognon (1979). "Périodes arides du Quaternaire récent et déplacement des sables éoliens dans les Ziban (Algérie)." *Revue de Géographie Physique et de Géologie Dynamique* 21 (2), pp. 97–108.

Bettis III, A. (2012). " Climatic and Biotic Controls on Silt Production and Accumulation of Loess ". *Nature Education Knowledge* 3 (10), p. 25.

Bisson, J. (2003). *Mythes et réalité d'un désert convoité : le Sahara.* Paris: L'Harmattan. 480 pp.

Bisson, J. and Y. Callot (1990). "le hommes et la sécheresse autour du Grand Erg Occidental." *Sécheresse* 1 (2), pp. 124–133.

Dejouhanet, L. (2006). "Le chemin de fer en Mauritanie - composant d'un modèle minier ou moyen de développement d'une région ?" *Le chemin de fer en Afrique.* Ed. by J.-L. Chaléard and C. Chanson-Jabeur. Paris: Karthala, PRODIG, SEDET, pp. 293–306.

Dubief, J. (1959-1963). *Le climat du Sahara.* 2 vols. Special isse. Algiers : Institut de recherches sahariennes. 312 & 275.

NOAA (1935). The "Black Sunday" dust storm of 14 April 1935. Available at: http://www.srh.noaa.gov/oun/?n=events-19350414-maps.

Oulerhi, T. (1992). *Étude géodynamique des migrations de sable éolien dans la région de Laayoune (Nord du Sahara marocain)* . Mémoires des Sciences de la Terre 92-12. Université Pierre et Marie Curie (Paris 6). 223 pp.

Rognon, P. (1994). *Biographie d'un désert.* Paris: L'Harmattan. 350 pp.

# 3

# Living in Deserts

*There are peoples who live by marching through the desert, of various origins: Berber, Arabic, black or mixed-race Africans; those who are light-skinned and those who are dark-skinned. Chaambas to the north, Peuls to the south, Moors to the west, Toubous to the east, Tuaregs in the center, spread out over three thousand kilometers from Algeria to Libya, from Mali to Niger, people with different languages and cultures, a vivid mosaic whose unique features are rarely seen by tourists, who are so used to looking upon Africa with a reductive gaze. You will find different crowds from Nouakchott to Agadez and Ghat, even if the stirring of the urban melting-pot, the movement of populations fleeing civil wars or drought and Western influences on lifestyle tend to blur the lines. There are differences not only in the styles of clothing with varying degrees of looseness or slits, but in the colors: the black, puffed-out sirwals of the Moorish and Tuareg camel-drivers, the tight, white trousers worn by the Toubous, the azure-blue tunics with their discreetly embroidered necklines of the Mauritanians of noble descent – light-skinned, aquiline featured, with fine-framed glasses – overseeing date harvests from afar.*

*There are people who live in the desert, whose differences are sometimes less striking than their similarities. Gracious, courageous people, dividing the rights and duties of men and women less unjustly than in other regions. A man is always away, taking risks in distant expeditions, on the lookout for meager resources, a hunter, herder, seasonal laborer, occasional warrior, leaving his non-reclusive wife to manage the finances, the tented domain, the family, the children, domestic animals, the tiny kitchen garden. A woman reigns over her domain, rarely accepting polygamy, sometimes initiating divorce, keeping her own property, autonomous.*

*All of them live a harsh life, their bodies worn by time, their eyes burnt by the sunlight, their teeth poorly cared-for and skin cracked, buffeted by the sand and wind. Some practice extreme frugality: flour, salt, sugar and tea. It is said that a Toubou lives on one date every three days, eating the skin the first day, the pulp the second and the seed on the third...*

---

Chapter written by Marc CÔTE.

*The region sees no obesity, there is a natural asceticism, an aristocratic bearing. These are people with the innate elegance of dress, of their striking hairstyles, of clothes fluttering in the breeze...*

Pierre Gilloire, *Itinérance au désert* (Itinerant Life in the Desert), Éditions du Rocher, 2001.

These "people of the desert" have a few things to teach us today.

## 3.1. Humans and the desert

### 3.1.1. *A "desert civilization"*

Robert Montagne was the first to use this expression, thus emphasizing the unity of life across this large collection of lands extending from the Atlantic to Central Asia and China. Studied in a holistic manner on the eve of this last century, these deserts appear to offer rich glimpses into a unique, original and thriving human life. This refers not only to a population or its activities, but also an elaborate cultural life. This is therefore a true civilization, whose roots can be traced back to the Neolithic revolution, that continues to grow now.

One cannot help being astonished by the continuity of life through deserts, the desire for survival among the population here or their ability to adapt to an environment with such significant constraints. Nothing prepares us for the fact that these vast spaces, blazing in the summer and freezing in the winter, bear so many different traces in the ground from the human societies that have inhabited them.

We must realize that, without deserts, humanity would have been deprived of many of the riches that awe us today: the architecture of the ksour, the beauty of the oases, the proliferation of markets and caravan journeys, the poetry and music of the deserts and the originality of urban life.

### 3.1.2. *There are deserts and then there are deserts*

A human presence is not found in all deserts, but is restricted to only a few of them. Humans have had a profound impact on the history of the Old World, on horseback in Africa and Asia, from the Sahara to the Chinese deserts, from Mauritania to the Gobi. This is sometimes referred to as the Arid Diagonal (Figure 3.1), which cuts across 15,000 km, from the shores of the Atlantic in the west to the Asian mountains in the east. All across this region we see, to varying degrees, the interwoven communities of oasis civilizations, Bedouin groups and urban life.

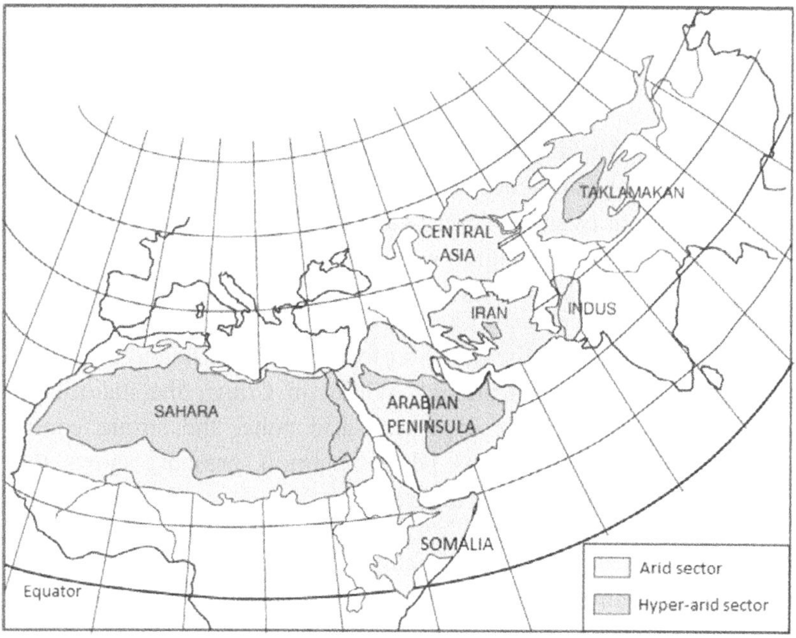


**Figure □.□.** *The Arid Diagonal within the ancient world*

However, in the other deserts that checker the planet – those in California, South America, southern Africa and Australia – we find neither oases nor Bedouins. These desert societies remained at an elementary stage of hunter-gathering.

A point that must be noted is that the desert does not automatically produce oases or pastoralism, as was clearly observed by Robert Capot-Rey. These phenomena are a response to human society, wherever in history we find them[1].

---

1 Note by P. Bourrié: Recent work by North American anthropologists have shown that the indigenous societies of the Americas also had to develop water management. We find oases throughout the cultural super area of Oasisamerica: Fremont, Pataya, Anasazi, Hohokam, and Mogollón, which cover the greater part of Utah, Arizona and New Mexico, large parts of Colorado, Sonora and Chihuahua, as well as minor extensions in California and Texas. There, despite the dry climate, the indigenous populations sometimes obtained two harvests a year (López Austin and López Luján 201□, p.□□).

The emphasis here will be on the spaces along the Arid Diagonal, which were almost exclusively the domain of these lifestyles. It was not till recently, i.e. the current era, that life in the deserts of the New World came into being and paralleled that of the Old World.

### 3.1.3. *The desert, Islam and oil*

While the Old World was marked extensively by the presence of deserts, this was not the only influence in the region. A large and vigorous cultural influence spread out over the desert, carried by nomads, travelers and cities: this was Islam. While this religion was marked by its desert origins, in return, it also often left its marks on these territories.

The general overlap of the borders of the desert world and the area of the Muslim world may be misleading. While, in China, the majority of the Muslims in the country live in a semi-arid zone, the strong contrast in Southeast Asia (tropical regions where Islam is present) shows that the relation between the two, aridity and religion, is not one of direct causality.

Just as this superimposition is complex in nature, so also is the recent relationship between this desert area and hydrocarbon-rich regions a matter of chance (Figure 3.2). It turns out that the Arid Diagonal corresponds to a geological domain with sedimentary basins that are rich in oil and gas. It is one of the most hydrocarbon-rich regions in the world, representing over 60% of world reserves.

This fortuitous occurrence has had certain consequences. Over the course of the past century, it has given desert states limitless economic resources and created entirely new landscapes, from Hassi Messaoud to Bakou, while, in the Gulf countries, it led to the adoption of lifestyles that were alien to these spaces.

All this has also had an impact on the desert. Thus, we must be careful not to attribute phenomena to deserts when they may actually have their origins in Islam or the Oil Age.

### 3.1.4. *Antagonistic and complementary civilizations*

There are drastic constraints in the desert. Many areas do not receive more than 100 mm of rain, while hyper-arid pockets may receive no rain at all for 5 years at a time. The desert sees torrid summer heat, frosty winter nights, high daily and seasonal variations, the absence of natural vegetation and severe dust storms.

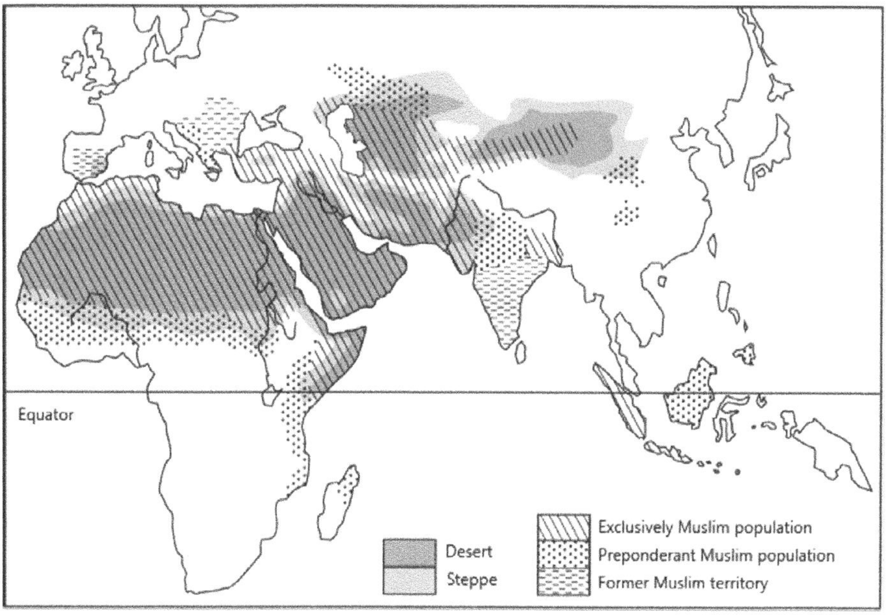


**Figure 3.2.** *Area covered by the desert and area where Islam spread and was established*

Humans can respond in a variety of ways to these harsh constraints. The simplest response would be the absence of any human presence. This has been the case in most deserts on the planet. However, in the deserts of the Arid Diagonal, the millennia-long history of living in the deserts and increasing demographic pressures has led to populations settling into the present environment and creating different ways of living in the deserts.

These ways of living differ greatly from site to site. There is an ocean of difference between an oasis, whose greenery is restricted to a space that is extremely limited, and the mobility of a nomad group traveling hundreds of kilometers. This diversity is, however, juxtaposed within the same territory.

One way of living here was, therefore, a specific, limited adaptation to the environment, based on spread, mobility and the intermediary of livestock. Over the centuries, and on a variety of scales and through very different modalities, nomad populations brought these spaces to life and founded the basis of a *Bedouin civilization*.

However, a different process was also possible: this consisted of going against the desert, seeking to combat the aridity by using the smaller, available water resources to create an "oasis effect" and generate a compact agro-system capable of protecting the populations from the effects of the physical constraints. Symbolized by the oasis, and repeated hundreds of times, this gave rise to a *hydraulic civilization*.

History also responded with a third form of organization. In order to give greater value to the vast area, and compensate for the scarcity of agricultural resources, some societies brought all their efforts to bear on building relationships, basing their economy on trade and establishing communication networks. Thus, they focused on all the nodes in the network, turning their cities into cultural hubs. They thus produced an *urban civilization* in the heart of the desert.

Therefore Robert Montagne's desert civilization can be divided into three different forms: three kinds of societies, each developing its own culture, producing specific civilizations. Three forms of life that were not inherent to the desert, but today make up the uniqueness of the Arid Diagonal. They reveal an astonishing creativity. The same desert has "generated" both oases and pastoralism, the camel trails and the establishment of urban life. The relationships between nature and society thus seem much more complex than they appear, even in the desert environment.

These three civilizations were all born in the same region, juxtaposed and often interwoven. There may often be tensions between them, but they are also complementary in their political, economic and territorial organization. The nomad groups, the oases and the cities have often lived related lives, with a common history and close-knit relationships, all of which created and breathed life into the vast territory of the Arid Diagonal.

## 3.2. Bedouin civilizations

### 3.2.1. *Nomads*

The Bedouin – a term used chiefly in the Middle East – are the very essence of nomads. That is, individuals who respond to the vast spaces and the aridity of the land through mobility, a flexibility in habitation, and through the role played by livestock. For the Bedouin, the vast spaces are not a handicap but, instead, an advantage that they can play on to compensate for the aridity of individual sites.

Their nomadism may have different purposes, for example, nomadic warriors are groups of men, generally on horseback, who wage war, conquer a territory, or pillage an oasis or another nomadic group. Mobility and rapid action are their chief assets.

Throughout history, the Sahara and the Arabian Peninsula have seen many small groups of Bedouins wage war, either to establish their group's supremacy or as the basis of their economy. In the Mongolian steppes, warrior groups on horseback built up vast empires.

Commercial nomadism may be an offshoot of the warrior groups, but the chief objective here is to establish economic relationships between cities or countries. The symbolic image associated with this commercial nomadism is that of the caravan trade, which, for a large part of the Middle Ages, brought alive the Sahara, along the gold routes, as well as the Asian territories lying along the Silk Road. Nomad groups would provide transport animals, guides and the protection for the caravan. Even today, dates and salt are traded through caravans and caravan routes (called Azalaï) in order to bring these products to markets along the fringes of the Sahel.

Finally, nomadism may also be carried out for pastoralism, i.e. the livestock are no longer simply an instrument for transport but form the basis of the group's economy. The nomad group's movements are governed by the grazing needs of their animals. This is why we speak of pastoralism (mobility of the livestock), which is a specific lifestyle under nomadism (mobility of the humans). Pastoralism is, by far, the most widespread form seen in the deserts.

### 3.2.2. *The power of the group*

The Bedouin civilization has a significant sociological dimension. The phenomenon of a community, which also exists in other civilizations, is at its most powerful in this civilization.

Many terms are used for the basic group: tribe, clan, faction, *douar*, *arch*, etc. The group claims a common ancestor and is led by a patriarch or group of elders, and is held together by links of blood rather than land.

There is solidarity against the oases and the villages as well as against other nomadic groups. This solidarity may take the form of building binary alliances between two groups (soff).

The names of some groups have gone down in history (the Moors, the Kunta, the Chaamba, the Banu Hilal, the Shammar, etc.). Some groups speak Arabic, and some Berber; some are white, and others are black.

The tribes of the desert and semi-deserts posed significant problems to Roman troops along the Saharan borders and Chinese armies along the Mongolian borders. They played an important political role in the face of colonization by foreign powers. The large tribes in the Arabian peninsula resisted this for a century, leading to the creation of the Saudi kingdom. The Bedouin tribes in the Sinai continue to resist and hold on to their territories even today.

The concept of belonging to a group may operate at very different levels. The Tuaregs are organized into a vast confederation numbering over 500,000 members, arranged into a hierarchy of groups and sub-groups. On the other hand, the Toubou people of the Tibesti ("the Black nomads") are relatively individualist and function in small groups closer to the family level.

### 3.2.3. *Astonishing knowledge*

The Bedouins were able to adapt to their environment because of their double expertise on the territorial as well as pastoral front.

Their knowledge of the territory is related to the space, the region and the area it spans. Their advanced geographical knowledge also allows them to read the signs in the landscape. They know places, names and potential resources across vast distances. Each massif, each wadi and, sometimes, each tree has a unique character.

This is how, over the centuries, the Bedouins have become guides *par excellence*, for their families, herds and tribes, as well as for tourists. It is their understanding of the land that allowed them to carry out warrior raids over 1,000 km: in recent decades, there have been raids by the Front Polisario in Mauritania, and raids on their Chad adversaries in Njamena.

This expertise is complemented by their expertise in pastoral activities.

The nomads may have been called "herd followers", criticized for carrying out a very elementary activity. In reality, however, pastoral rearing of livestock is a highly developed lifestyle. The *in situ* experiences mean that the nomads have intimate knowledge of the life of their animals, of their foraging, the sub-products, illnesses, their reproduction, the paths to take each season and the behavior of the herds throughout the year. This knowledge has allowed the nomads to live for centuries in the steppes or arid environments without any harmful effects on the environment. Problems with ecological imbalances have only manifested over the last century.

### 3.2.4. *The fundamentals of pastoralism*

Given that pastoralism is the major form of nomadism, some of its fundamentals must be described.

First, we must discuss the ecological aspect: the animals are adapted to the environment. The herds most often consist of goats and sheep, complemented by a few camels that ensure the logistical efficiency of the camps.

The movement of the herds makes up for the low biomass available: a daily displacement over a short distance and seasonal displacements over larger distances, all of which take into account green zones, the edible species in every season, and the presence of water points (Figures 3.3 and 3.4).

In addition to this, some groups also carry out annual movements, moving out of their usual areas. This is called transhumance, and this becomes very significant when the topography has mountains and basins (e.g. the mountains of □entral Asia and the Saharan Atlas and within the Tibesti).

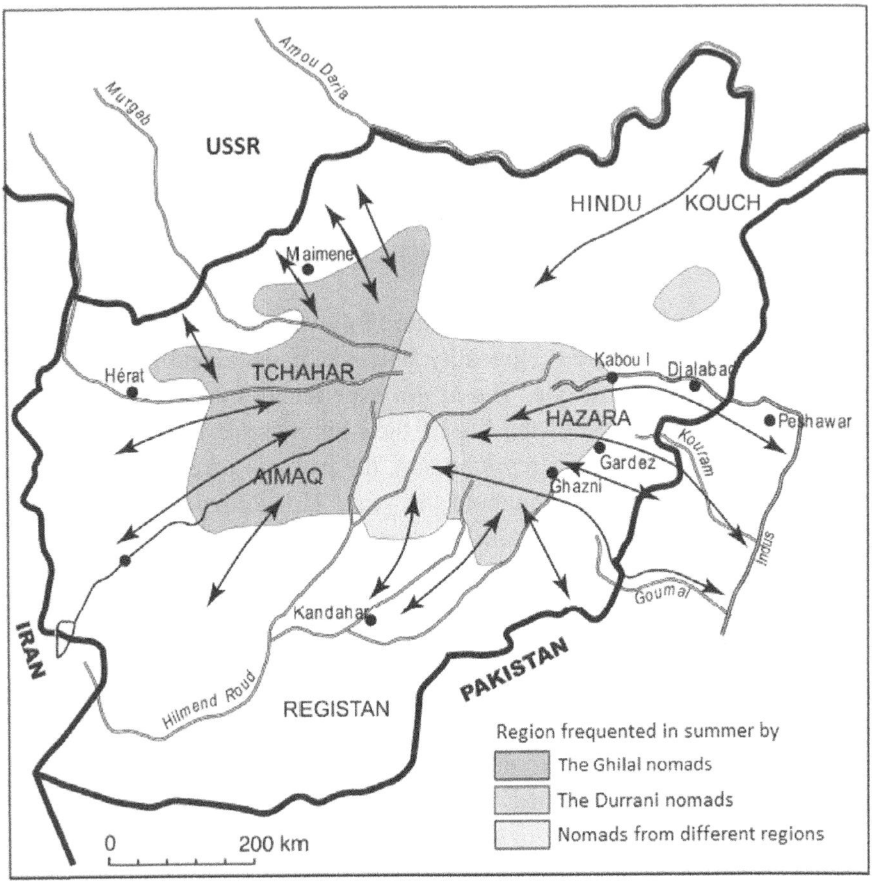


**Figure □.□.** *Transhumance in the mountains of Afghanistan*

There is also a strong sociological aspect to this practice. The herds belong to each family, but the paths are collectively managed, as it is the group that decides new paths to explore, organizes the camps, monitors the grazing (in order to avoid over□grazing) and establishes contracts with other groups for the right of passage.

The logistical complement of this kind of mobility are tents and camp sites, symbols of a lifestyle that has no geographic fixity.

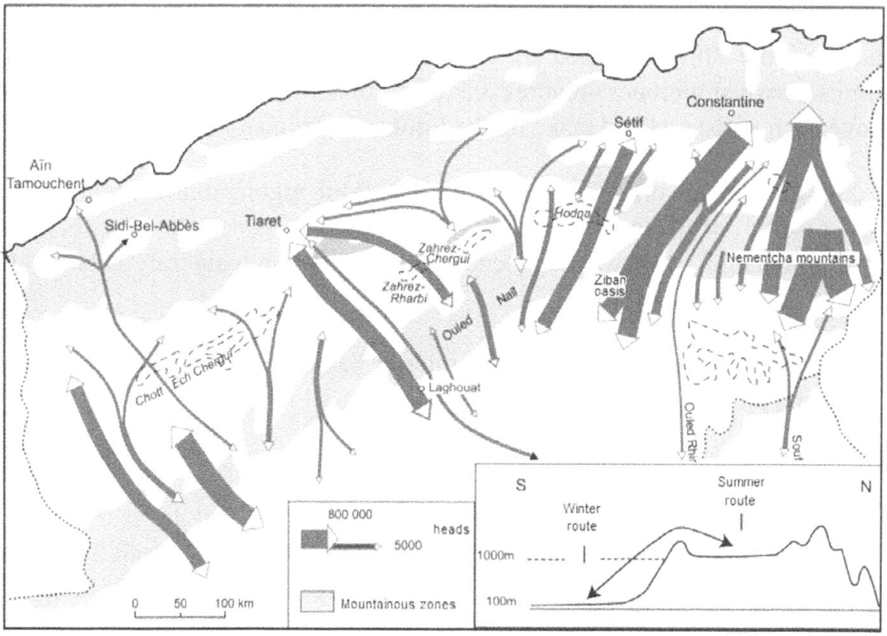


**Figure □.□.** *Transhumance in the Saharan Atlas*

### □.□.□. *Nomad territories*

This term may appear paradoxical, however, nomadic pastoralism is based on an imperceptible, but rigorous, spatial organization of the territory.

In order to survive in the long term, the group must be able to master the territory. Thus, they tend to organize the territory over long distances and long durations and to get other groups to acknowledge this.

The desert is, therefore, spatially organized, based on the interplay between ecological terroirs and the tribe's borders. For instance, the mountainous massifs (in the Saharan Atlas, in the Near East, in Yemen, in the Zagros Mountains, etc.) have biotopes in parallel bands (mountain, slope, piedmont, plain), and pastoral groups have mapped their territories[2] so as to optimize the complementary resources found within the biotopes.

2 Note by □. Bourri□: Specifically, the territory used by a particular rural community is called a "finage", from the Latin □□□□□ limit.

This resulted in the classic configuration of long terroirs, the borders between the "finages" used by different tribal groups, cutting across the limits between biotopes (Figures 3.5–3.7). In some countries, these borders have been registered and make up the limits of administrative communes.

In this framework, the groups benefit from regular transhumance and are assured of their grazing rights. They are generally described as "semi-nomadic". Some groups tend to settle down in temporary habitations (made of plants or clay).

The "large-range nomads" of the interior Sahara or the Arabian peninsula are much more subject to ecological vagaries.

### 3.2.6. *Collective lands*

The same piedmonts that border deserts (which the Arabs call *Sahel*, or shores) benefit from another asset: the presence of cropland, cultivated by Bedouins.

The arid lands in the piedmont can receive inflows from floods in the wadis that descend from the mountains situated inland. Through a system of mini-dams, earthen levees and large canals, the people here ensure the floodwaters spread out and help them flow in the form of a "directed inundation", in the hope of being able to harvest grain. This is how several nomad groups were able to lay claim to these floodplains (the various names given to these lands indicate how important they are to all the tribes here: *grair*, *djelf*, *faïd*, *maader*, *henchira*, etc.). Another sign of their importance is how this property is treated: the lands belong to the entire tribe, are shared equally between members and are redistributed every 5 or 6 years by drawing lots.

This mode of sharing, which has been passed down through the ages, has persisted for a long time and is still practiced in certain areas (for example, the Nemencha lands, in eastern Algeria). This highlights the great influence of the community system.

In other regions, collective lands were decollectivized, either by public powers (as in Tunisia, Syria and Mongolia) or through decisions taken by tribe leaders, as a result of pressure from their own people.

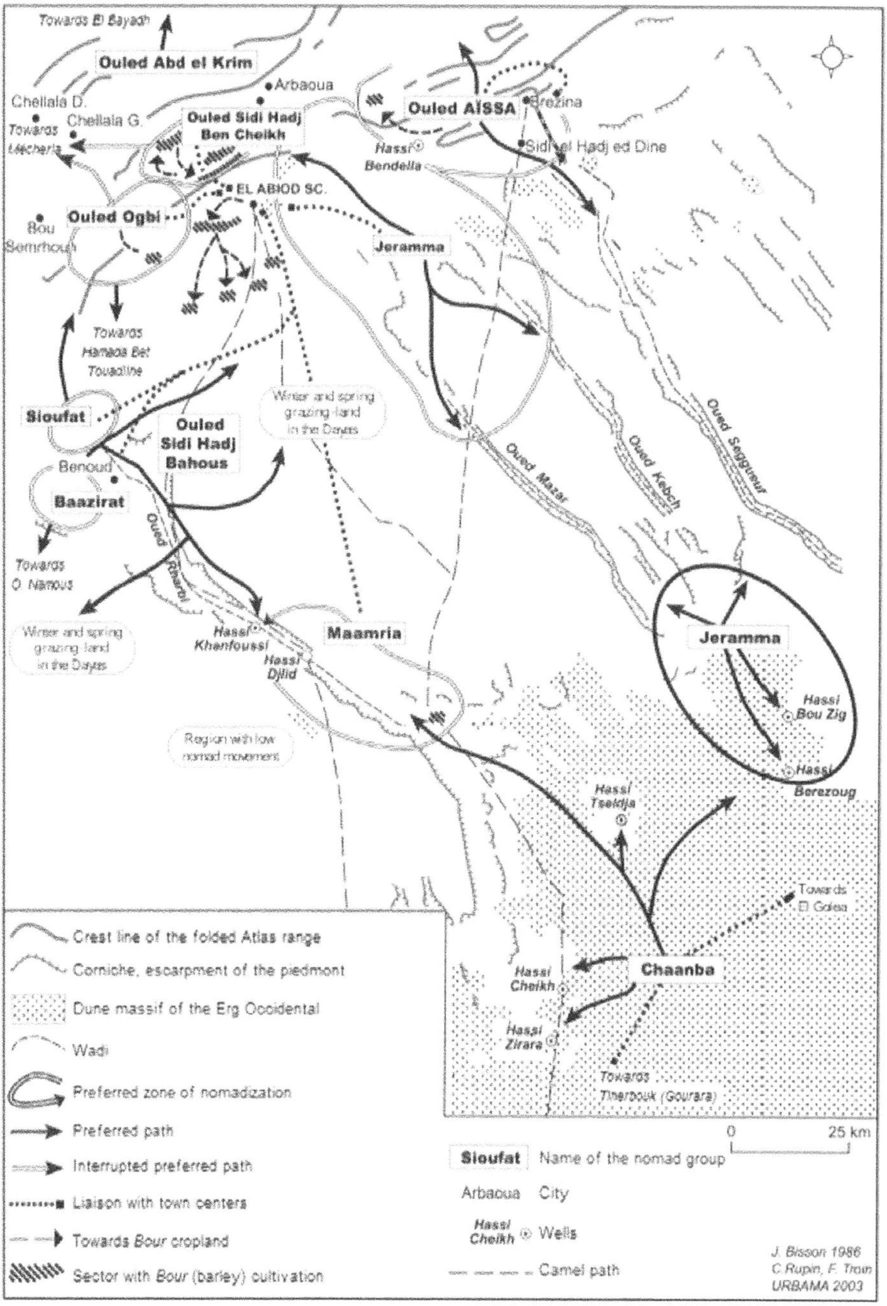


**Figure □.□.** *Pastoralism over small distances. The area traveled by the Ouled Sidi Cheikh (western Algeria). Jean Bisson speaks of "casanier nomadism" or "stay-at-home" nomadism*

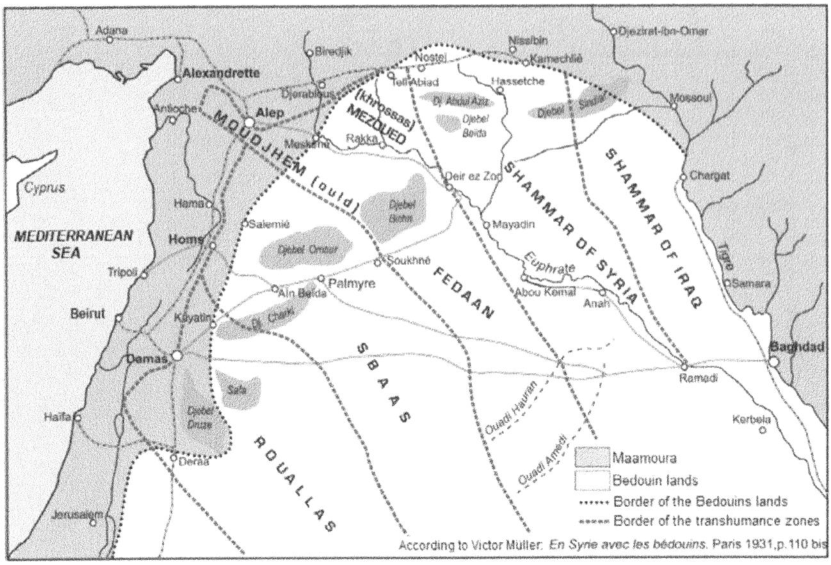


**Figure □.□.** *Nomad territories in Syria. Spaces where transhumance is practiced on horseback over the Maamoura (cultivated lands), the Djebels, semi-arid spaces and arid spaces*

### □.□.7. Dromedaries and Bactrian camels

The deserts along the Arid Diagonal vary in appearances.

They also vary in habitation. The best-known form is that of tents made from woven cloth, with each nomad group using a different color. However, other groups use huts made from braided straw (Chad), yurts (Mongolia) or constructions made from plant material or clay (regions in the Sahel).

The deserts also vary in the animals used for transport and production: dromedaries, Bactrian camels, horses, yaks, cows and buffalo.

Schematically, the deserts along the Arid Diagonal can be divided into two groups. The west (the Sahara and Arabia) is a region with vast lands, nomads who travel large distances and the dromedary. The dromedary is a camel with a single hump and is an animal that has proven to be extraordinarily well adapted to the environment. A draught animal, as well as one used for its meat, it can live in large herds, or travel alone carrying a Bedouin. In this western part of the Arid □iagonal, there are at least 17 million dromedaries. This region has hot deserts, situated along the Tropic of □ancer.

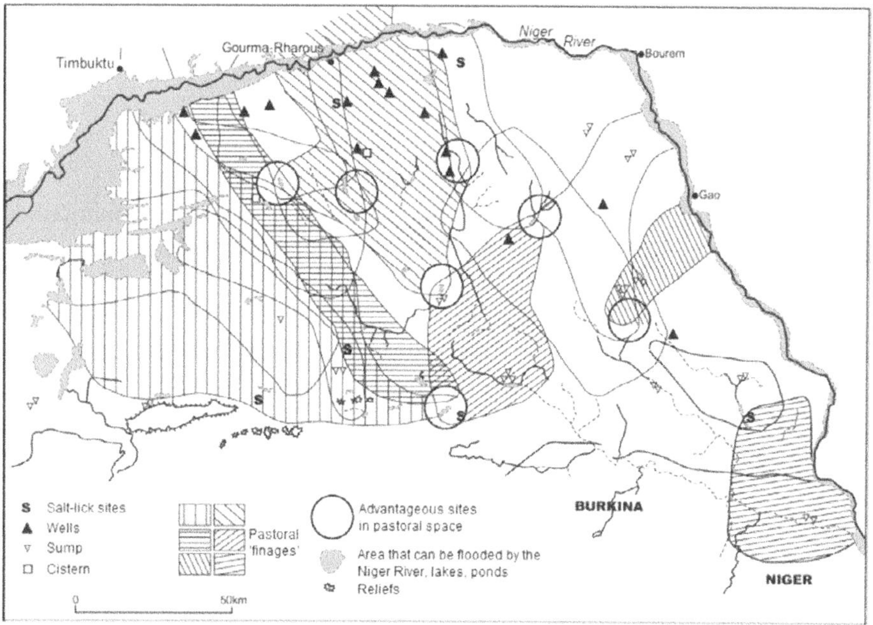


**Figure □.7.** *Nomad territories in the Malian Sahel, in South Gourma, in the loop of the Niger river. Finages of different nomad groups stem from the shores of the wadi, up to the meridional plateaus, making use of the complementary features of each biotope (Gallais 1967)*

The east (□entral Asia) is the region of the Bactrian camel (a camel with two humps). This is smaller than a dromedary, is robust and can tackle slopes, rocks and the cold. The higher altitudes and higher latitudes along which these regions are located make this the region of cold deserts. □nly the Bactrian camels can be used as transport in this environment.

There are about 1.5 million Bactrian camels, up to the □obi □esert.

The paths of the Bactrian camels from the east and the dromedaries from the west crossed along the Silk Road, leading to the birth of more resistant animals.

X. de Planhol contrasts Arabian nomadism and the nomadism of the Turks, observing that the former is based on the use of dromedaries and has seen the conquest of large areas, did not penetrate much into the mountains and remained largely sedentary, while the latter used Bactrian camels, practiced transhumance and saw much greater penetration into mountainous regions.

### 3.2.8. *A large-scale ecological transformation: desertification*

This human–animal–environment balance has not always been maintained. It has been progressively disrupted over the 19th and 20th centuries.

The chief reason for this was the excessive population of animals with respect to the paths along which they traveled. On a local scale, this was manifested by "aureoles" of desertification, forming around each grazing well. On a regional level, this was reflected devastatingly by the decade-long drought from 1970 to 1980 in the Sahel, which saw horrifying levels of plant, animal and human mortality.

This excessive livestock load was worsened by the rapid extension of work carried out in the steppe zones (in Algeria, Turkey and Syria). Equipped with machines and wanting to plant more grain, the nomads cleared vast areas.

The alarm over the prospect of desertification was raised by international bodies. Studies carried out over recent decades have made it possible to respond to two essential questions: where had the phenomenon occurred? Who was responsible?

Investigations showed that the phenomenon was widespread but essentially concerned the steppes that bordered the deserts: the deserts could not really be further desertified.

While droughts and global warming may have played a role, the desertification here seemed to largely be an anthropic phenomenon, and the blades used to clear the land seemed to have caused more damage than the grazing animals.

Several administrative measures were put in place to limit the damage done. In Algeria, tillage along the steppes was banned. In Syria, the 200 mm/yr isohyet was fixed as the dividing line between the domain of the sedentary groups and the domain of the grazers (Sanlaville 2000).

### 3.2.9. *Social transformation: sedentarization*

This ecological fragility reflected the social fragility of the pastoral world. There would have been no ecological degradation without human changes.

The chief cause for this was sedentarization. Any State in the world mistrusts nomadic life. Their objective becomes to control the movements of the nomads, pin them down to a region and find them habitation, i.e. sedentarize them. Not a single colonial or nation state in recent decades has proven the exception to this rule.

Nomad groups have often responded to these efforts out of a desire for modernization, more comfortable habitations and formal education for their children (Figure 3.8).

Thus, this movement has been quite generalized and sedentarization has most often been accompanied by impoverishment. This phenomenon was most marked at either end of the Arid Diagonal: in Mauritania, where the severe droughts led to the nomad population falling from 70 to 5%, and in the Mongolian steppes, where state pressure led to widespread sedentarization of the population into urban complexes.

During the colonial period, political borders were slapped down on nomad territories. The territories thus became fragmented and the nomads' displacements became curtailed (Figure 3.9).

The logical consequence of sedentarization and territorial control was a weakening of intra-group alliances. Thus, individualism progressively spread across the steppes.

### 3.2.10. *New forms of pastoralism*

One would assume that pastoralism would stagnate, stunned by these various blows. However, this was not the case. The nomads sought to adapt to modern life, just as they had adapted to the biotopes of the Middle Ages.

Some of these adaptations were as follows:

□carrying out a reconnaissance of the paths, after the rains, using portable all□terrain vehicles□

□mechanizing the transport of animals, on trucks, which allowed them to travel across the regions quickly and without tiring□

□traveling backwards along the path to set up food reserves and water tanks□

□finally, settling the women and children into cities while the head of the family set out with the herds.

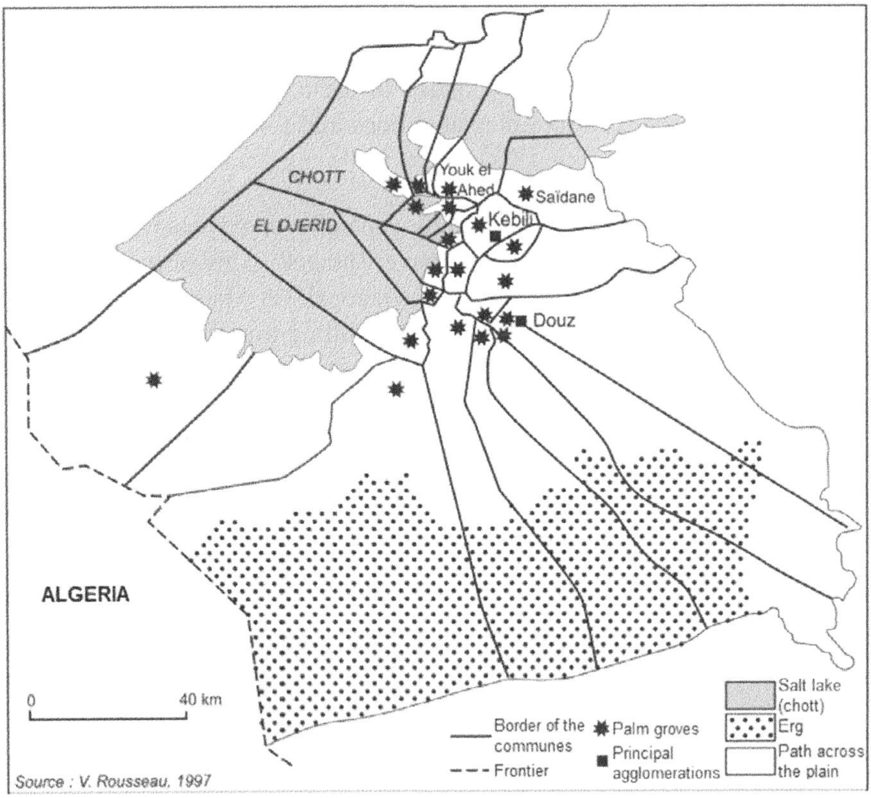


**Figure □.□.** *Nomad territories in southern Tunisia. The territories have been divided such that each section includes palm groves, chotts, plains and ergs. Current communes use older borders (Rousseaux 2004)*

This evolution thus resulted in a dissociation between the family and the nomad shepherd and the social elevation of mechanized livestock□rearing at the expense of the small, sedentary livestock breeders. □astoralism (the movement of animals) continues, while nomadism (the movement of humans) disappears. □hat will happen with the next generation□

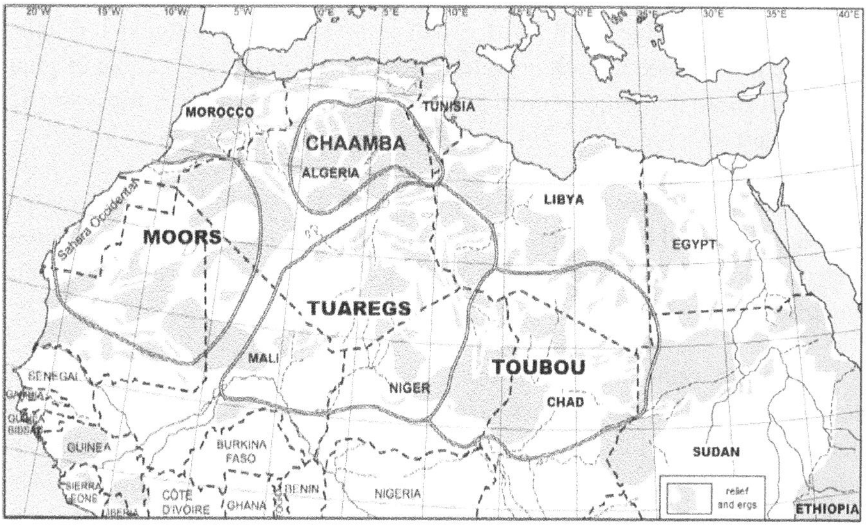


**Figure □.□.** *Tribal borders and political frontiers. Territories of the four large nomad groups in the Sahara. Political frontiers established during the Colonial era. Each group found its lands fragmented (C□te 2014)*

### □.□.□. □□e □e□ □a□es o□□ro□□ str□□t□re

□hat then remains of the groups□

In countries that were extensively colonized, the new administrative framework that was established caused the features of ancient groups to disappear. □owever, in semi□arid and arid regions, historical alliances and blood ties were much better maintained than was believed.

A symptomatic sign is: local political life, the issues around elections and the existence of associations all led to a resurgence in tribal affiliations, the role of large families, the functioning of rural zaouias and fraternities. This phenomenon was quite evident in the interior regions of the □aghreb, in Libya, in Arabia and in Afghanistan, where these factors played an active role in political life.

Thus, life in the desert held out against time.

## 3.3. Hydraulic civilization

### 3.3.1. *Oasis life*

An oasis – an Egyptian term – can be very simply defined through a convergence of three factors: a water resource, a habitat and an area of green cover. All these combine to create a dark spot in the middle of the ochre sea of the desert.

The water resource may be a spring, a well or a wadi; the habitat may be a village protected by ramparts or in the middle of a palm grove. The green stretch is most often a palm grove, but may also be an olive grove or a poplar plantation.

The basic principle in the oasis is the interdependence of the three elements and their compactness. The high density of trees and their increase in size from the periphery to the center ensure that the water, habitation and delicate agriculture, all of which are located in the center, are protected. This is the "oasis effect", which leads to a considerable decrease in heat and evapotranspiration in the oasis.

Thus, across vast and open stretches, the oasis is a point, a contracted space that is created so as to effectively resist adversities.

### 3.3.2. *The fundamentals of hydraulic civilizations*

A hydraulic civilization is an oasis multiplied by $n$.

Foreign observers have always been stunned by this persistence that led to the creation of pockets of life in areas with such harsh conditions. Was this not luxury agriculture? Was this not a strange form of living?

In fact, the oases were not born as agro-systems. They were sites for life to emerge, relays, for a relational life that needed logistical support. Caravans required points of support at intervals, if possible, regular intervals, where they would find water, food, fodder and accommodation. The oases were created to respond to these needs, built up by physical labor wherever a little water existed. Sovereigns, war lords and tribal chiefs all undertook to create oases along a route in order to support it.

The site of an oasis was decided by the possibility of accessing water, but they were first created to set up relay posts along caravan routes.

### 3.3.3. *Water territories*

It is possible to define three kinds of hydraulic territories.

The first – which we just saw – is the simplest: the oasis, properly speaking. This is an isolated site, remote from any other oasis, which had to be created, at all costs, to disenclave a region. Because of its vast size, the Sahara has led to the largest number of isolated oases: Ghat, Ghadames, El Menia, Tabelbala, Kufra, Siwa and so on.

However, these oases are often grouped closely together in space, forming clusters of oases. Over a 30–60 km radius, one finds tens of oases packed as closely together as possible, with each cluster being as distant from another as possible.

Identical geomorphological and hydraulic conditions, affiliation with the same human community, a desire to build up logistical relay posts for long caravan journeys: all these explain the existence of oasis clusters that mark arid regions. Examples of such clusters include: Touat, Gourara, Saoura, M'zab, the Oued Righ, Nefzaoua, Fezzan, Borkou and the Kaouar.

Finally, there is a third type of hydraulic territory: large hydraulic groups that have made such a strong impact on certain plains of the Arid Diagonal that one forgets these are plains in an arid zone. These zones are on an altogether different scale, whether spatially or temporally, and have been created for other purposes, a fact that must be emphasized (Figure 3.10).

### 3.3.4. *Cradles of hydraulic civilizations*

Egypt, Mesopotamia, Central Asia, the Indus Valley, and Northern China: all these are large civilizations with hydraulic irrigation, cradles of civilization that were prominent in the ancient world for millennia.

| System | Elevation in meters | Type of energy | Number of hectares irrigated per day |
|---|---|---|---|
| Archimedes Screw 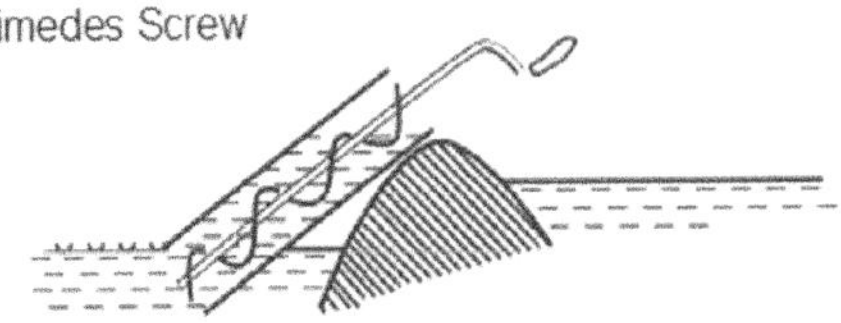 | 1m | 1 man | 0.3 ha |
| Shaduf 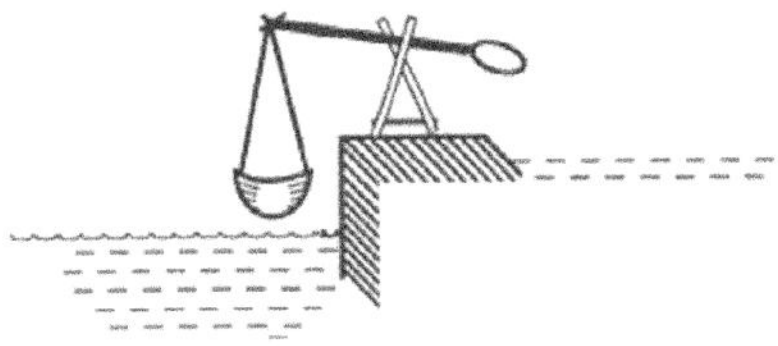 | 1m | 1 man | 0.1 - 0.3 ha |
| Noria  | 1m | Power supply | > 0.8 ha |
| Saqiyah 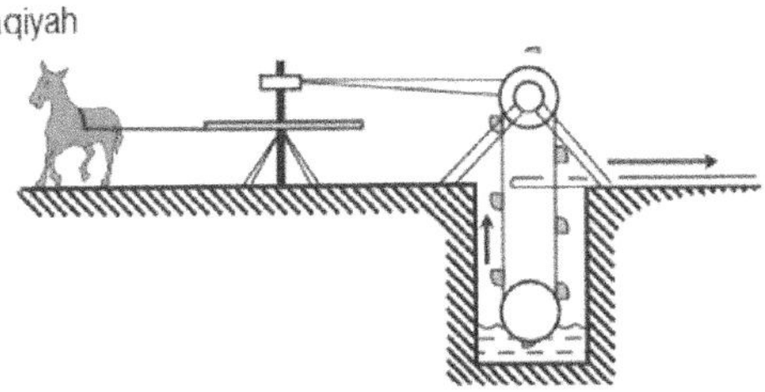 | 1m | Animal traction | 2 - 4.8 ha |

**Figure □.□□.** *Traditional water-lifting systems: Archimedes screw, shaduf, noria (water wheel), saqiyah. The information given for each system: elevation in meters, the type of energy and the number of hectares irrigated per day (Heathcote 1983)*

The large, allogenic rivers (the □ile, the Tigris and □uphrates, the Indus and the □ellow River (□uang □e)), which were difficult to control, because of their floods, but which had remarkable hydraulic power, were the earliest to justify large water infrastructures. The scale of the large dams, networks of canals and the area under irrigation were all vastly different from those resulting from oasis clusters.

Recent hydraulic forms and techniques can be tracked back to ancient systems: Egyptian, Persian, Roman, Arab, Chinese and so on. Architects consider infrastructures like the Egyptian storage structures, the Ma'rib dam and its sluice-gate in Yemen, the large canals of Central Asia (Samarkand), the vast palm groves in the Chott el Arab and the Dujiangyan dam in the Chinese Chengdu valley to be gems of the golden age of the hydraulic civilization. Egypt, Mesopotamia and the Indus Valley were centers of hydraulic civilization that spread across a large part of the Arid Diagonal. The large modern dams are remodeled versions of the dams of these three civilizations. However, these large-scale projects were thwarted by land hunger in Egypt, large flooding in the Punjab region and cross-border tensions in Mesopotamia (Iraq/Syria/Turkey) (Figures 3.11 and 3.12).

The creation of the Saharan oases required the intervention of local powers (war chiefs, the important patrons of the caravans). *A fortiori*, these significant installations on large water bodies implied a close-knit relationship with a strong, centralized power. This has long been demonstrated in Egypt as well as in China.

On the southern flank of the Sahara, the Sahelian band did not see any large hydraulic areas, despite the presence of the Senegal and Niger rivers, Lake Chad and the Upper Nile. This region only saw a few mediocre oases and harvests from palm groves. The African kingdoms in these regions in the Middle Ages, Ghana, Mali and Songhai, certainly had a state structure, however, they never built up a "hydraulic network". Even today, their hydraulic installations are on a modest scale.

### 3.3.5. *Hydraulic techniques*

Desert populations were able to perfect different forms of accessing water, depending on geomorphological conditions as well as their own technological advances (Figure 3.13).

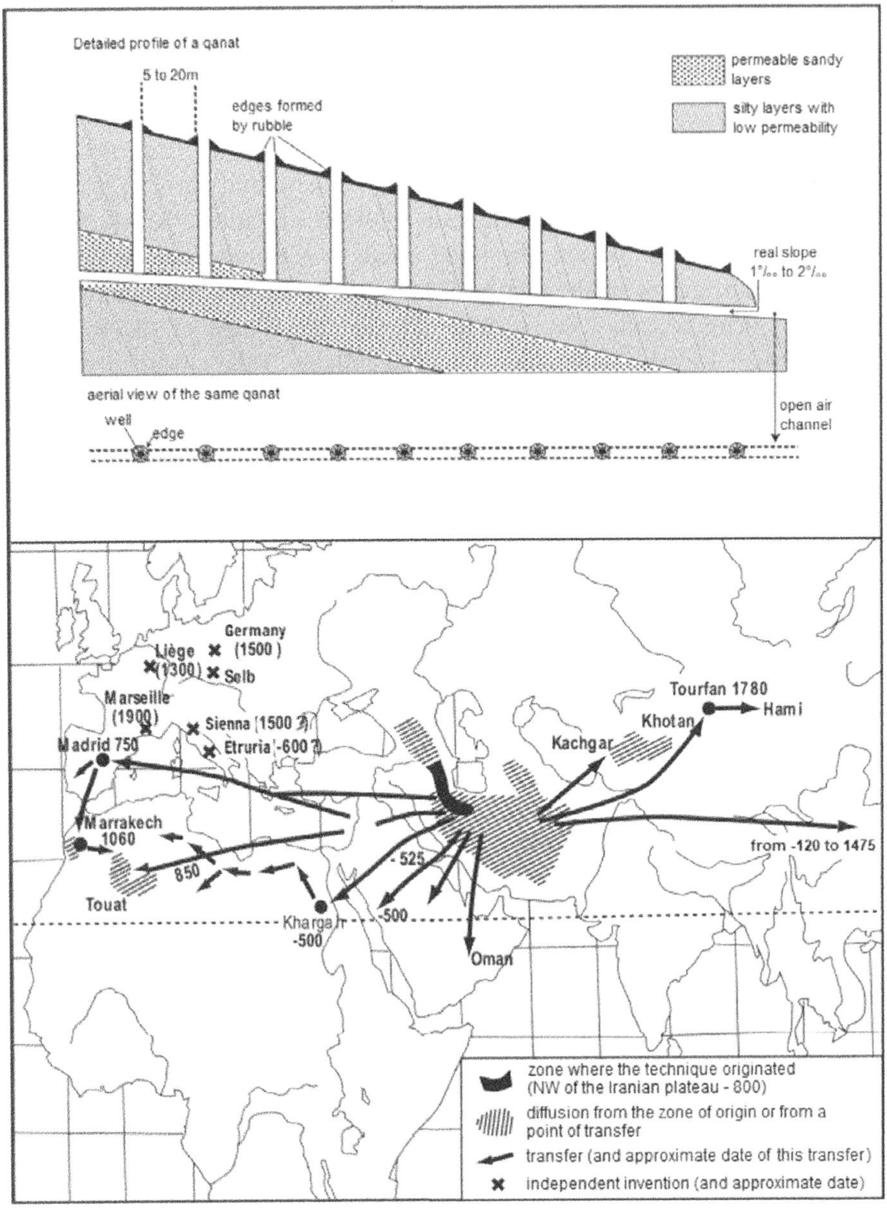


**Figure □.□□.** *The foggara technique. Technique for underwater drainage of a water table. The success of this technique in Iranian territory has led to it spreading toward China and, notably, toward the Sahara, where it remains in use today in western Algeria (Goblot 1979)*

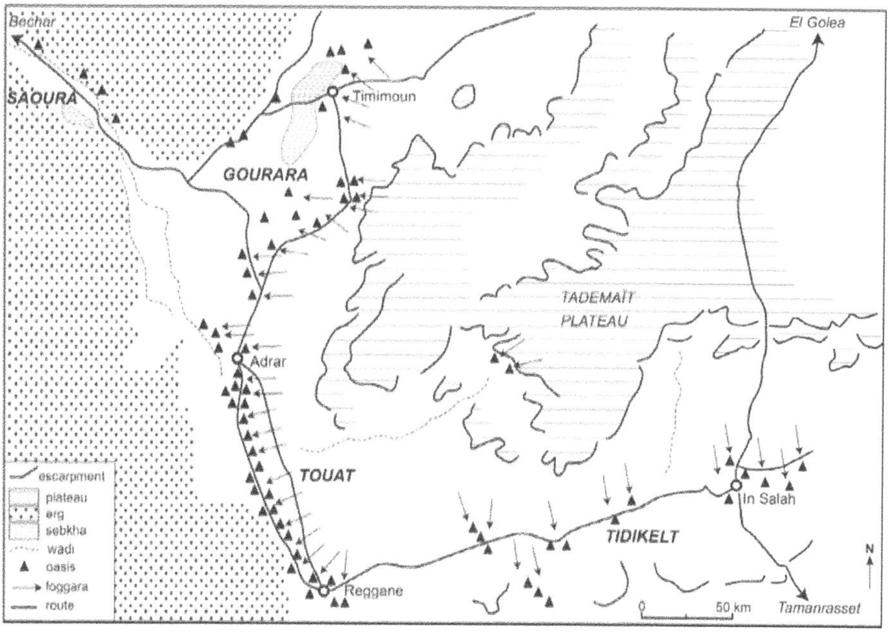


**Figure □.□□.** *An example of the Saharan hydraulic systems. Gourara, Touat, Tidikelt; 1,500 foggaras around Tademaït (Côte 2002)*

The first kind involves the capture of superficial water (springs, wadis). It would appear that desert populations very quickly learnt how to construct dams (made up of earth, stone and plant matter) and dig canals of all sizes (some are tens of kilometers long).

When there is no surface water, ground water must be accessed. Thus, many populations focused on ways to raise water from the ground. They first had to dig wells (and this was not a simple operation, some of these wells reached depths of □0 or even 80 m), then line the well, and then finally raise the water. Wells with a balancing wheel, noria, wells with buckets, wells with a water screw and wells with a pulley system: these were the many different kinds of wells that marked desert landscapes. These were basic techniques, only allowing small areas to be irrigated (less than 1 hectare).

A unique form used was that of capturing water horizontally, through underground channels: foggaras. The foggara would drain the aquifer through a slight slope that would meet the aquifer upstream, but was small enough to reach the surface downstream. □xcavated wells, breaking through the soil on the surface, were used to extract subsoil materials. The foggaras involved significant work initially but then ensured irrigation was carried out through simple gravity. They were known by many different names across the □iagonal (*□a□at*, *□ar□□*, *□□attara*, etc.) (□igure 3.1□).

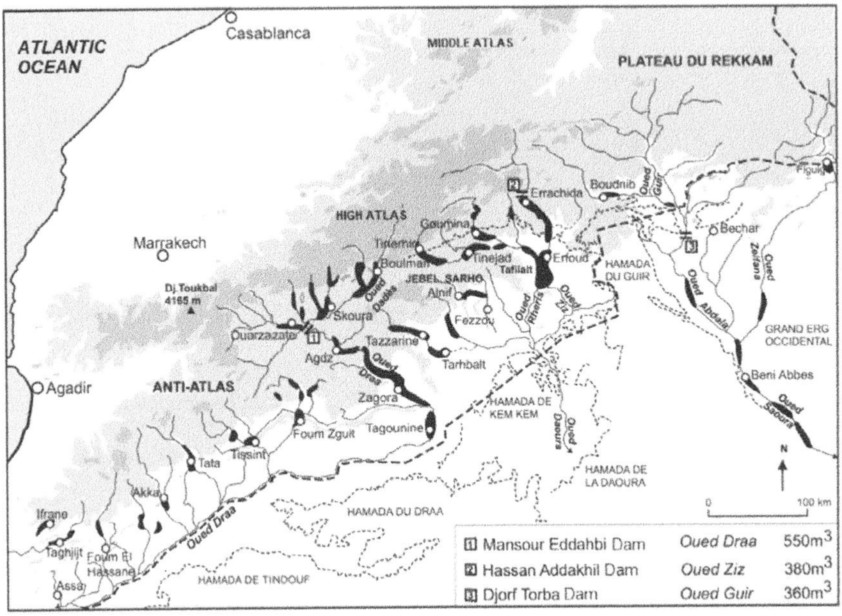


**Figure □.□□.** *An example of a Saharan hydraulic system: the oases in southern Morocco on the piedmonts of the Anti-Atlas. Oasis with valleys and gorges, renovated by modern dams (two in Morocco, one in Algeria) (C□te 2002)*

### □.□.□. □o□e□ti□e a□d i□di□id□a□□□dra□□i□ s□stems

This is perhaps the most significant difference between hydraulic systems as it has fundamental implications from a societal point of view.

It appears that wherever individual hydraulic systems were possible, they were implemented. This was because they were based upon an elementary, family unit: a well with a water□wheel, a well with a pulley, a noria, all corresponding to individual land use.

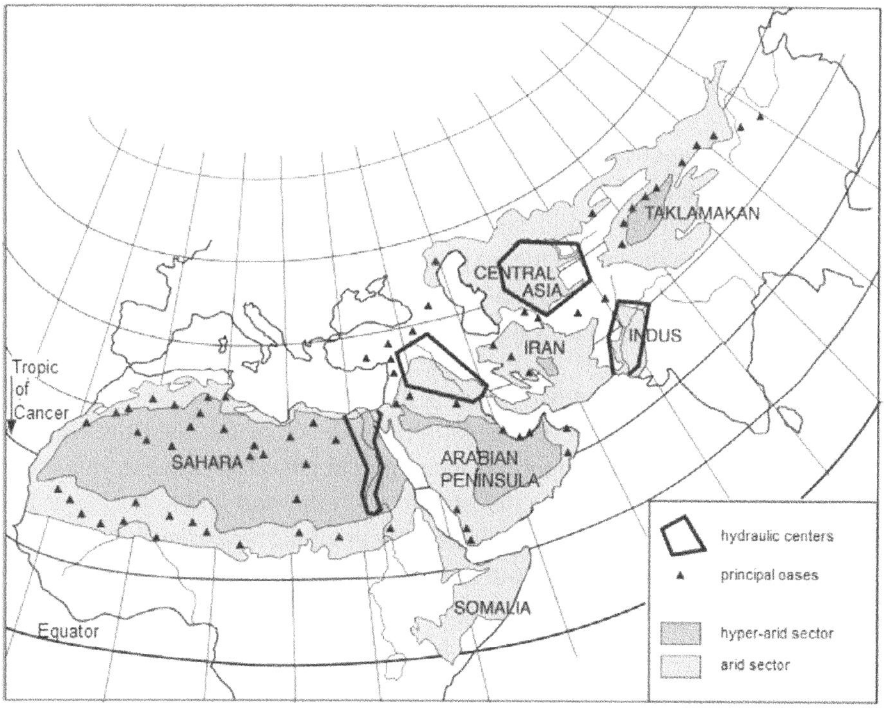


**Figure □.□□.** *Hydraulic systems in the Arid Diagonal. Hydraulic civilizations were born along the large points of convergence of water bodies.The oases are associated with a relay function (gold routes, spice routes, silk routes)*

□owever, water is difficult to access in the desert. It often involves pooled resources, the actions of a collective and the authority of a leader. The creation of foggaras, artesian wells and dams (even elementary dams) all involved collective action. □ollective action was also required for downstream processes: the maintenance of installations, the cleaning of drains, the sharing of water and the settlement of disputes.

Thus, the functioning of a hydraulic system involves collective action and authority, a social group in the oases, a State (as we have seen) in the large areas irrigated. □ydraulics, the State and collectivity mutually reinforce each other.

Today, the modern technology of a motor□pump and the rise of individualism in agriculture tends to give the first system primacy.

### 3.3.7. *Modes of sharing a rare resource*

Water sharing is the basic symbol of this supremacy of the collective.

A traditional practice, which marks the difference between an arid zone and rainfed cultivation, is the right to property. In rainfed cultivation, water goes with the land: when one is sold, so is the other. However, in an arid zone, water is dissociated from the land. One may have a right to water without having any land, or one may own land and still be obliged to lease water rights. Water is so important that the right to water has been individualized.

The technological genius of oasis-dwellers is reinforced by judicial prowess. An entire branch of legislation has developed around the right to water. This is especially important when a collective technique is involved, which presupposes a collective investment from the start and a sharing of the water when it arrives. The practice generally combines two means of sharing: sharing by volume, when the flow is large enough to be divided and to flow in canals; sharing by time, when the flow is limited and requires pooling together the entire volume of water for a certain time period for the same beneficiary. Thus, a large spring sees its volume divided into four channels that run through the year (each proportional to the number of right-holders); each of these canals, in turn, is divided into seven or 12 smaller canals from which the users divide the water, taking turns (1/6th or 1/12th the time for each user). An aiguadier (or water official) is designated to ensure that proper distribution is carried out. The law must be respected (but not necessarily justice).

As one may imagine, such a complex system can give rise to a number of disputes between rights-holders: in arid zones, conflicts over water are legion. Thus, we can understand why in sectors with large hydraulic networks there are entire "water legislations" and that certain sectors have "water registers" (Ghadames).

### 3.3.8. *The diffusion of techniques*

Deserts today are something of a museum for technologies developed to cope with an arid bioclimate (Figures 3.15 and 3.16). This leads to the question: how did these technologies spread over such a vast area?

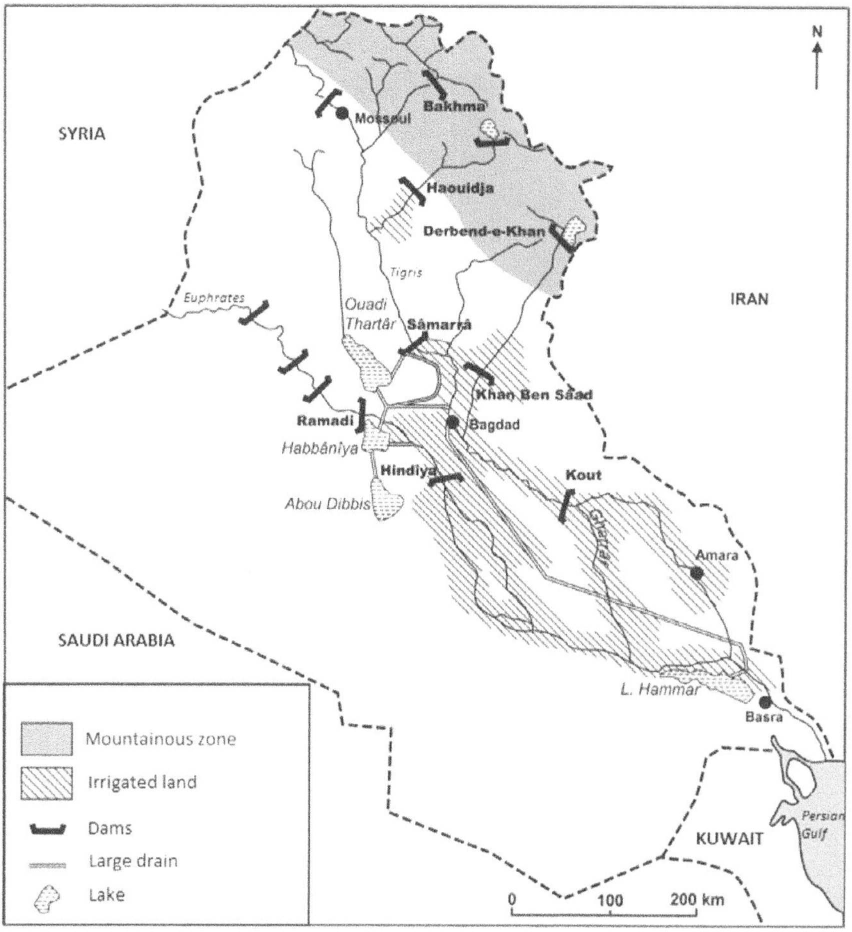


**Figure □.□□.** *The hydraulic system in Mesopotamia. These plains owe their rich soils and water supply to the Tigris and Euphrates. Dams regulate the waters and borders divide them*

One of the first ways is diffusion through proximity, from one center of innovation. This is how the technique of the foggara seems to have spread. They were developed in Persia, cradle of innovation *par excellence*, where the alluvium-rich piedmonts were perforated by thousands of drainage galleries (qanat). From here, we find remains of these wells in the Middle East, Egypt, Libya and up to western Algeria (Gourara, Touat and Tidikelt) where there are close to 1,500 wells (active or dead), which are today a part of the hydraulic heritage of the land.

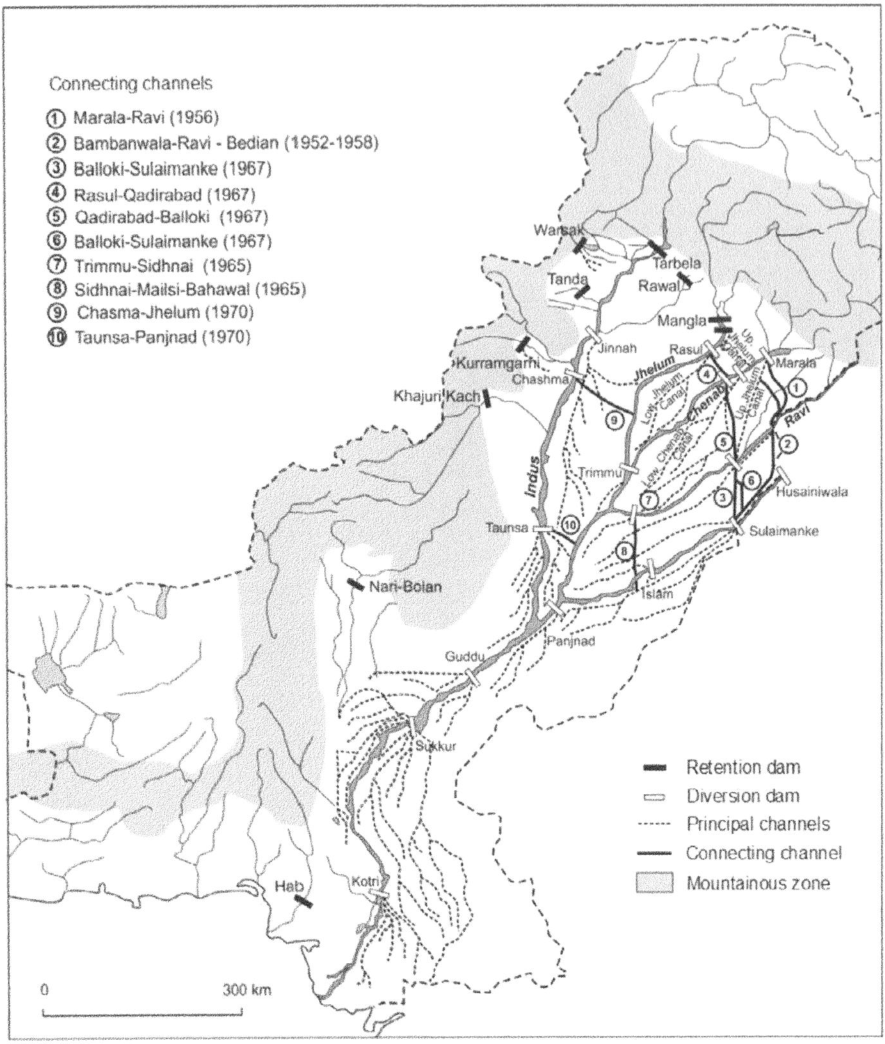


**Figure □.□□.** *The hydraulic system of the Indus Valley. "Today, the developed Indus regulates one of the largest and most spectacular irrigation systems in the world" (Durand-Dast□s and Mutin 1995)*

Another example of diffusion, with a dominant role played by an initiator, is that of the A□r valleys, where the action of a local, 18th□century Sheikh led to the replacement of water□wheel wells with wells with pulleys, driven by draught animals, which were thrice as effective, a small, local revolution.

However, another mode where a technique is adopted seems to result from a convergence of identical responses to the same environment and the same problem. This thus creates identical landscapes over thousands of kilometers.

Indeed, this seems to have been the case with the palm cultivations, generated progressively from natural palm groves and date seeds either discarded on the ground or planted deliberately. Today, these have led to landscapes of palms, as seen in the Mauritanian Adrar, Fezzan, Kawar, Borkou and Tassili. This is a slightly elementary technique on the borders of the Sahel, unlike the intensive mode used by the date palm cultivators on the northern edge of the Sahara.

### 3.3.9. *Hot deserts and cold deserts*

These analyses lead to the identification of differences within deserts. Schematically, the landscapes of the Arid Diagonal can be divided into two types, which coincide with the domains of the dromedary and the Bactrian camels. To the west (Sahara and Arabia), the landscape is largely made up of vast flat stretches on the geological basement of the African-Arabian plate. The absence of cold and dryness in the air is conducive to large groves ("forests") of palms. These are the hot deserts, situated on the Tropic of Cancer. They are made up of the visual triad of plains/palms/dromedaries.

The eastern part is more fragmented and rugged. There are higher altitudes and the region is located at higher latitudes, making this a land of cold deserts. The 40 °C temperature experienced in the summer may be matched by the −40 °C temperature experienced in the heart of winter. The palm cannot survive here and is replaced by chains of poplar (and small-scale farming, market-gardening) along the valleys and on the piedmonts. The Bactrian camel is the pack-animal used throughout these harsh regions. This world may be said to consist of the visual triad of valleys/poplars/Bactrian camels.

The limits of palms, olive-groves and Bactrian camels/dromedaries all define these well-characterized regional groups (Figure 3.17).

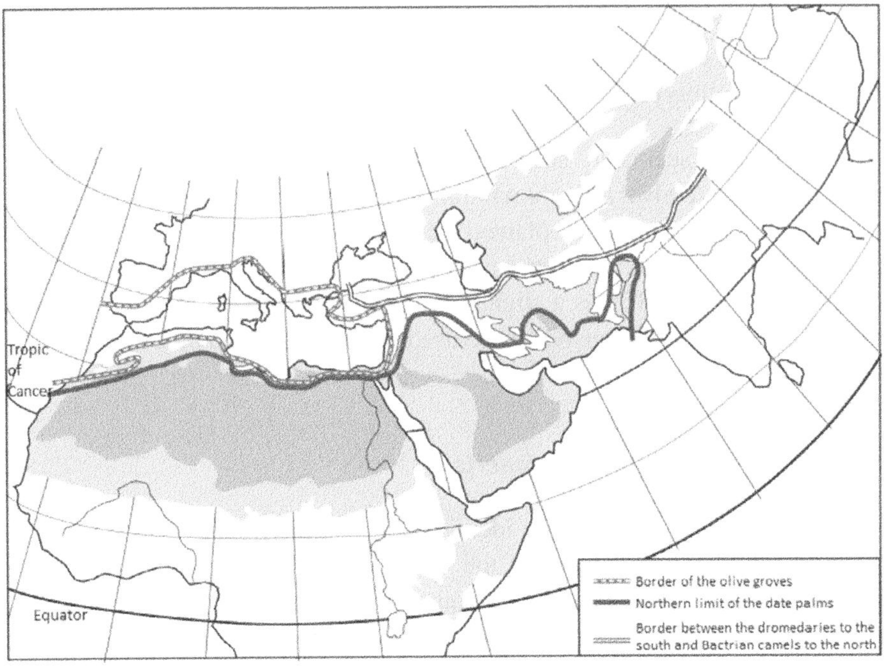


**Figure □.□7.** *Some fundamental limits in the deserts of Arid Diagonal. Limits of date palms, olive groves, dromedaries and Bactrian camels (Capot-Rey 1953; Hulum 1959; Bethemont 2000; Sanlaville 2000)*

### □.□.□□. *Today: from small wells to deep drilling*

The oases began with wells that were 5 to 10 m deep, with a small flow, which could only irrigate small areas. However, the 20th century saw the rise of new techniques that were introduced from the outside world: motor-pumps and vertical well pumps, which enabled drilling to moderate depths, drilling as deep as 1,000 or 2,000 m. Every time, the irrigation capacity of the machine would increase more than proportionally, going from a tenth of a hectare to 30 or even 100 hectares.

In parallel, public powers intervened in the development of the territory through the construction of masonry dams on the allochtonous water bodies of the large plains (Syria, Iraq, Indus Valley, etc.). These new techniques changed the amount of water that became available through hydraulic systems, leading to a change in the landscape and the actors.

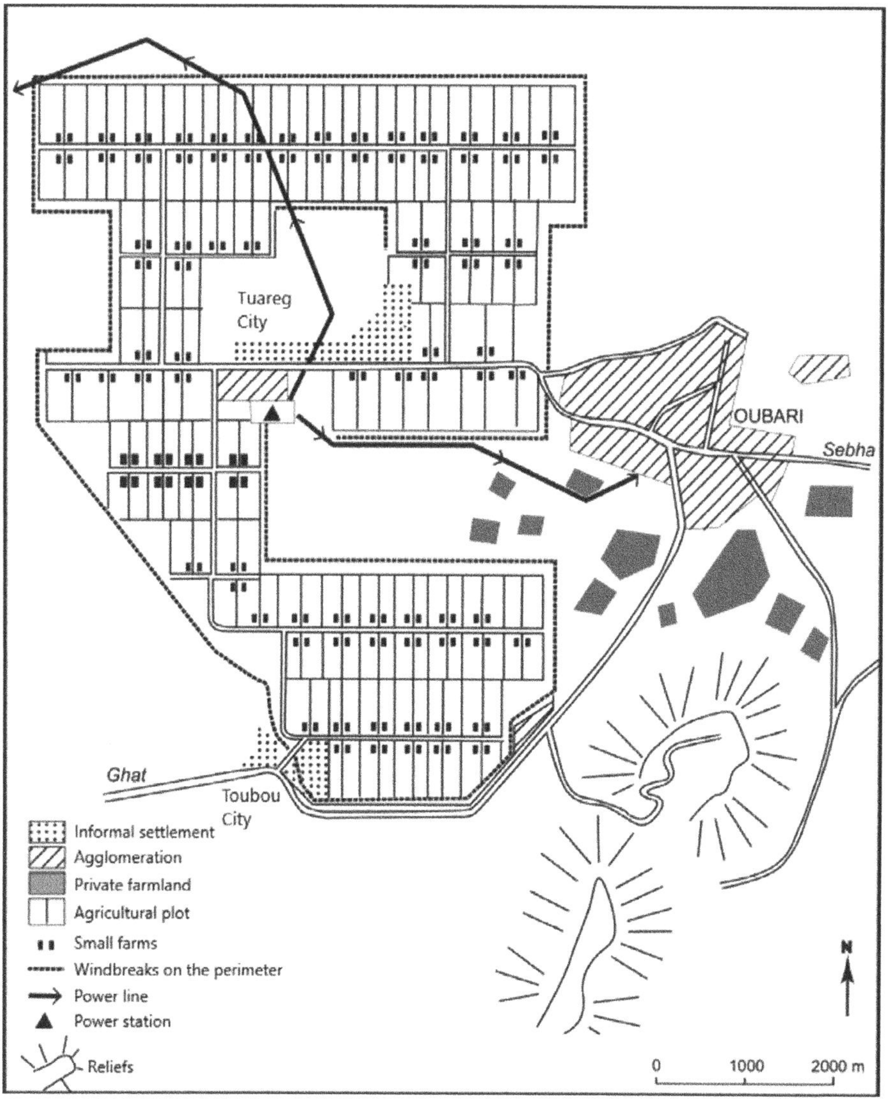


**Figure 3.18.** *State infrastructure and individual land-use. Perimeter of the agricultural land created during the 1980s; 1,800 ha, 10 ha/plot; close to the city of Oubari, Fezzan (Libya) (Côte 2004)*

Along their edges or further away, small oases saw the rise of large, irrigated perimeters, well-aligned, with fencing and plantations (Figure 3.18). The initiative for these planned plots was taken by the public powers who then allocated these plots to smallholders. The objective in this case was largely social: to provide employment and income to smallholders within the confines of their oasis. In parallel, however, in the context of the more liberalized economy of the past few decades, there emerged a more economic objective, which handed over agricultural land to largeholders, town-dwellers, liberalized professions or private companies. These large investors did not hesitate to use the most modern irrigation methods, such as agriculture in poly tunnels and pivot-irrigation for cereal crops.

Consequently, Algeria, Libya, Egypt, Syria and Arabia saw a large movement for agricultural development, which shook up the deserts with varying levels of success and concerns over deep aquifers. The enthusiasm for large-scale projects is today countered by the fear of significant ecological consequences.

## 3.4. A salt civilization?

Active societal life, ancient and recent, in the Sahara has often been linked to the presence of a geological resource, i.e. salt (see Volume 2, Chapter 4, "Salts").

### 3.4.1. *Societies in quest of salt*

Salt is present in large quantities in the soil and sub-soils of the Sahara in the form of mines, wells and springs.

Further, through a remarkable coincidence, it is seen that the Sahelian regions are cruelly lacking in this compound. South of the 14th parallel (the Niamey parallel), we enter an intermediate zone lacking terrestrial salt to the north (Sahara) and marine salt to the south (Gulf of Guinea). "There is thus a break between Saharan salts and marine salts" (Bernus 2006). However, societies know that mineral salt is a vital physiological requirement for humans. Thus, they have turned toward the deserts for many centuries.

Indeed, Monod (1937) noted "this deep hunger for salt which grips the peasant, whether in the savanna or in the forest, and which like livestock on the prairie, has produced and maintained one of the most powerful and oldest trade currencies there ever was, which is active even today" (Monod 1989, p. 174).

A hunger that through the ages has stimulated exchanges between regions, drawn out synergy between territories and given rise to tensions, sparked off ardent quests and led to high transaction costs.

Medieval Arab geographers (Ibn Battuta 1982) had this observation on the role of salt: "the Blacks use salt as money, as we do elsewhere with gold and silver; they salt into blocks and trade using this". A Portuguese explorer (V. Fernandez) noted in the 15th century, with respect to the African coast: "Kings offer more gold in exchange for salt than for any other product, they consume it themselves, as do their livestock, and claim that without salt neither they nor their herds could prosper".

Thus, over the centuries an entire geopolitical situation arose around salt on this part of the planet.

### 3.4.2. *Salt territories*

Salt was found along the median axis of the Saharan, between the 14th and 24th parallel. However, it is found in diverse forms and at a variety of sites. Progressive investigations have shown that usable sites form a large band, cutting across the Sahara, from the Atlantic to the Red Sea. This train of active sites can be followed from west to east.

On the Mauritanian coast, the Awjil salt works had the advantage of being closest to the Sudanese world and to the kingdom of Ghana. At the foot of the Ijil Iron mountain, the Idjil salt mines, which were probably active many centuries ago, saw a revival in the 14th century and were reputed for the quality of their product.

The Teghaza salt mine, in the heart of the Madjabat region, depended on Oualata and was coveted by Moroccan rulers, before it was supplanted by the Taoudeni salt mine. Both of these mines ensured Timbuktu's wealth and prosperity.

The Amadror salt mine, to the north of the Hoggar Massif, has long been mined by the Tuarags. Further south is the Tegguida mine, which was used both as a source of salt, on the salt cure route, and also as a copper mine.

The Kawar region was famous for the Segguedine mines (white salt) and the Fachi and Bilma mines (gray salt, for animal consumption).

The Borkou basin consisted of a series of small salt mines: Ouniange, Démi, and Bodo. The Démi mine supplied Ouaddaï Arabs from Abéché.

In Northern Sudan, volcanic lands were accompanied by groundwaters precipitating natron, mined by local populations for artisanal or medical purposes.

Hence, there were a variety of sites, only a few of which were used for commercial purposes, since it was difficult to bring together mining conditions found at other sites, or because their location meant that they offered no particular advantage. This was the case with the large Algerian-Tunisian chotts or salt lakes.

### 3.4.3. *The logistics of salt*

Salt was thus present across these territories and had to be inventoried, mined, brought to the surface, transported and sold; in short, all the logistics around this resource had to be put in place.

Each tribe or society perfected means of extracting salt, which were adapted to the nature of the deposits. Rock salt, for example, was extracted in plates (a camel load was made up of two to four plates). Surface salines (sabkhas or salty springs) were used to wash the salty terrain, decanting the salt water into small basins, then creating heaps of salt in different forms.

The second requirement was the creation of living spaces close to the saline, precarious habitations, villages or sometimes fortresses.

The third logistical element was setting up the caravans (made up of dromedaries) of varying sizes, from tens to thousands of heads.

The final requirement was tracing the routes for these caravans to follow, connecting a saline with a consumption zone. These itineraries naturally adopted a north-south path, connecting a saline in the north with a human community, or village, in the south.

Thus, we have the paths that are laid out schematically in orthogonal configurations. Indeed, the long and complex history of these relationships led to two different models of routes being established. The Taoudeni model is a path that forms a continuum from one side of the Sahara to the other, with a distance of about 2,000 km. The Kawar model, on the other hand, covers the southern half of the Sahara, as far as the Sahalian regions, covering a smaller distance of about 1,000 km. The flow of salt in this case was along a segment of the Saharan route and involved exchanging salt for local productions such as grains, thus establishing relations between nomads and villagers.

Downstream of these sites, salt was distributed through markets in the savannas and forests, through a hierarchic network that supplied the entire Sahelian region and established relations between all of these regions.

### 3.4.4. *An important period in the history of the Sahara*

Empirically and progressively, these exchanges generated an entire Saharan economy. The most spectacular example of this is the Kawar region, which saw the Azalaï (see salt caravan) travel south, fully loaded, exchanging salt for various local products, and these, in turn, for grains, carrying the bags of grain back toward the salt regions. Thus, a triangular trade was set up, ensuring a solid foundation for these populations. This region added copper works (Teggida), the production of dates (In Gall) and the "salt cures", which saw herds of animals traveling routes with salt deposits in the semi-arid zones for two months a year.

Another salt-based exchange led to new communities forming in a sector that was quite harsh, with the Teghaza salt deposits (15th century), subsequently replaced by the Taoudeni deposits. The rock salts mined here were transported by dromedaries over 700 km, as far as Timbuktu. The mining was carried out by Tuaregs and the trade by Berabish merchants. Every year, the Timbuktu (or Taoudeni/Timbuktu) Azalaï brought this Saharan route to life. The caravan would comprise of up to 40,000 dromedaris, transporting 140,000 salt bars.

The Bilma Azalaï transported around 3,500 tons of salt every year, while the Timbuktu Azalaï transported 5,000 tons of salt.

In every region (Mauritania, Kawar, Borkou, etc.), this trade had significant regional consequences. It gave rise to agricultural life, villages and urban centers such as Timbuktu, Gao, Agadez, Abéché and so on.

### 3.4.5. *A salt civilization?*

Through this long-distance commerce, which has been carried out over a long period of time, the history of salt became an integral part of the more general history of Saharan trade and the large caravans connecting the Mediterranean and the Sahel. Slaves and gold were traded, from south to north, in exchange for arms and cloth from north to south. These caravans kept the region alive for over 10 centuries (9th–19th). Over all or part of the region, the salt trade flow led to the doubling of routes and relay posts. In popular imagination, the salt trade led to the doubling of the gold trade.

However, the activity of the first has rarely affected the magnitude of the second. The rulers of the Sudanese kingdoms and the large Berber merchants of the Sahel did not find as much wealth in salt as they did in the gold trade.

During the Timbuktu Azalaï, the Kawar Azalaï or the salt cures, the Sahel witnessed festive scenes. However, we must also recall the darker side of this activity. The bitter life in the salt made the hyper-arid conditions twice as harsh. Houses built of salt blocks (Taoudeni) and the remoteness of these sites from any other inhabited regions, as well as the absence of oases, led to people speaking of the "salt hell". Salt gave rise to economic activity, but very rarely to life.

Thus, the role of salt mining as a driving force in the southern Sahel regions slowly diminished. The volumes transported kept increasing well into the 19th century (it attained its peak toward 1960, in Timbuktu) but the Azalaïs now faced competition from trucks, while the political events of the last few decades led to economic exchanges faltering. Kawar alone has managed to retain a certain volume of trade.

Having already been doomed as a factor of social and economic life, it has long since vanished as a foundation for a civilization.

## 3.5. Urban civilizations in the desert

### 3.5.1. *Cities*

It may seem paradoxical to speak of cities in the middle of the desert. The phenomenon we are looking at here is that of a town, but it is also urban as it carries a connotation of more elaborate social organization, a richer cultural life and a more rigorous planning of space. It is a connotation of greater advances in the implementation of architectural and urban structures.

Torn between the "bedouin life and the urban life" (Berque 1958) the city in the desert has features that differentiate it from a ksour (village), but also differentiate it from nomad tribes, creating veritable medinas (cities). The city in a desert is a major center, spreading out over tens or hundreds of kilometers and it organizes, commands and ensures the application of the law, dominating a network of surrounding settlements.

A desert city owes this urban quality to two factors: the long, historical tradition of urban life in these regions (stretching back over several millennia), and the privilege accorded to a city in a Muslim civilization (which extends across the major part of the desert).

### 3.5.2. *The fundamentals of urban life*

The city is the site of artisanal activity, manufacturing and trade. Oases and nomad tribes lack the space required to carry out artisanal activities. Thus, these activities are carried out in cities (Figure 3.19). The weaving of carpets and burnous, pottery, wickerwork, woodwork, making copperware, working with leather and goldsmithing are carried out, either at home or in stalls, or again in specialized neighborhoods. This is especially the case with blacksmiths, who form a caste apart in most cities.

The city is also a site for intra-urban trade (elementary consumer goods) and trade from distant places (gold, slaves, dates, salt, etc.). It is the city that manages and sponsors this large-scale commerce from distant regions, the large caravans carrying rare products. In order to fund this activity, the city counts on its bankers, money changers, bazaars, fondouks, caravanserais and warehouses of all kinds. In the desert, where agriculture is only a secondary activity, trade is the major activity.

The city is also a center of power. This is where we find the casbah (fortress) the bordjs (fortified structures), the administration and the seat of the representative of central power. The city tends to impose its will on the oases as well as on the tribes, which often organize themselves to form a counter□power. The city continues to exercise dominance over the region and collects revenue from the countryside.

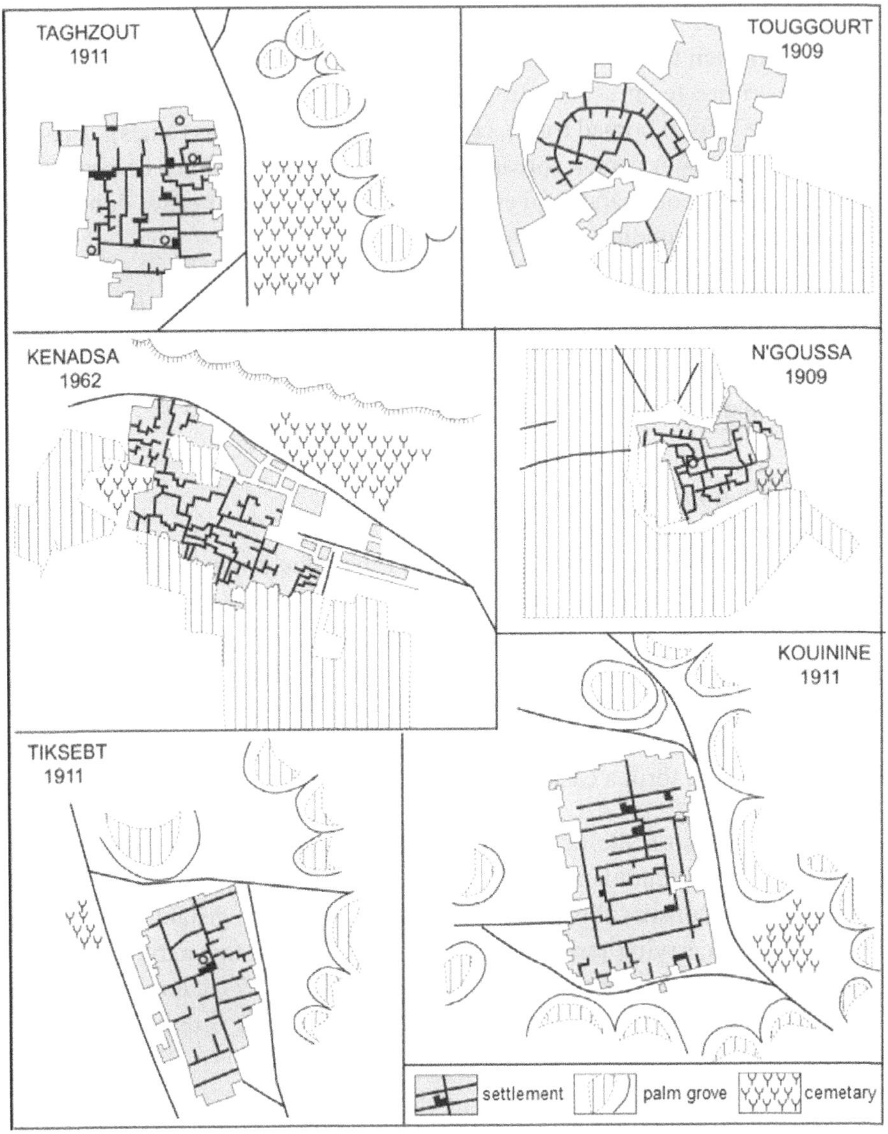


**Figure □.□□.** *Saharan architecture. Some examples of a ksour, large compact burgs, enclosed within palm groves and cemeteries. Images from the Algerian Sahara (Echallier 1968)*

Thus, the city establishes relations with the world of the oases and nomads. It is a synthesis of Bedouin civilization and hydraulic civilization. As de Planhol and Rognon (1970) stated, “in other words, we need a city in the desert”.

### □.□.□. *The city and the road*

The roads have given rise to cities in the desert. Desert cities have given rise to roads. This is a particularly strong symbiotic relationship in the desert.

Below, we discuss a few typical examples of these routes, where humans needed to cross the desert and thus created cities there.

For many centuries, the Silk Route connected two large cradles of life, far-removed from one another: Central China and the Middle East. It was a commercial route (which saw the trade of silk, paper, porcelain and fabric); it was also a pilgrimage route (along the path of the Buddha) and gave rise to political constructions (the Mongol kingdoms). Snaking across □,000 km, it presented a single route, which sometimes doubled back on itself (from one side of the Takla Makan desert to the other). Oases and cities emerged along this route. Some of these have died out today, while others are still alive and thriving (Figure 3.20).

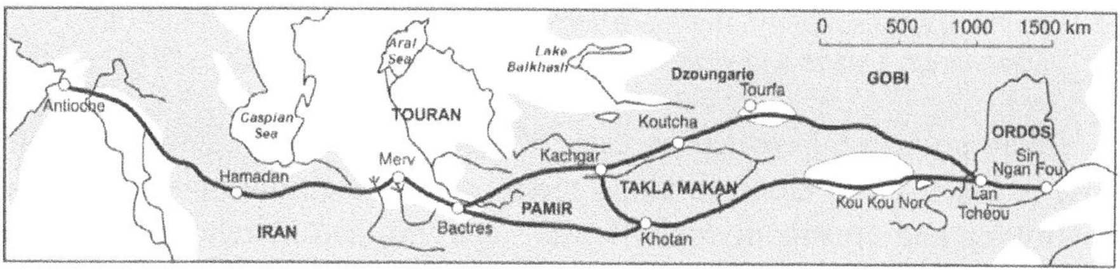


**Figure 3.20.** *The Silk Route, a route that connected countries from the Levant to interior China, across* □,000 km, *across the deserts of Central Asia. Today, it is punctuated by dead or living cities (Demangeot and Bernus 2001)*

This large route was active from the beginning of our era until the 15th century, when it faced competition from sea trade (the Spice Route).

We find a Spice Route (incense) in the Arabian Peninsula. Through the Middle Ages, caravans would transport incense, from the regions of *□ra□a □e□□* (Yemen) to the septentrional cities and Mesopotamia.

A third example is the □□old Routes□, which snaked across the Sahara, linking the northern and southern edges of this vast desert, facilitating the trade of arms and fabric from north to south, while gold and slaves were traded from south to north. The sprawling length of the desert (5,000 km) led to the emergence of seven or eight such □□S routes, which competed with each other (□igure 3.21).

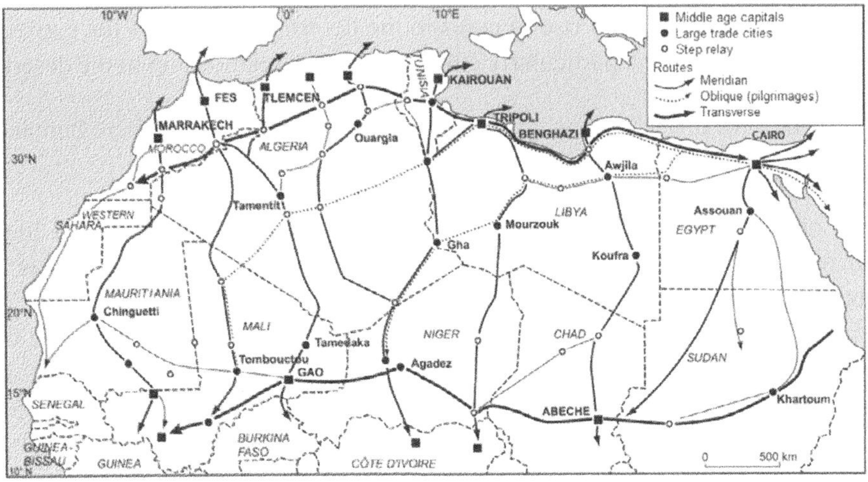


**Figure □.□□.** *The Gold Routes. The caravan trails between the northern and southern edges of the Sahara were in operation between the 10th and 19th centuries. Oases and villages emerged along these roads (C□te 2014)*

Relay posts were created on each of these trails, which slowly became oases. Later, cities emerged along these oases, which took on the role of controlling and organizing trade flows. □ach trail had its relays. The most important of these were the gates into the desert: the northern gates in the □aghrib (Sidjlmassa, □uargla, □ab□s, Tripoli, etc.) and the southern entry points in the Sahelian region (Aoudaghost, Timbuktu, Tadekka, Agadez, etc.).

Trade needed the desert. The desert gave rise to cities.

### □.□.□. *Net□or□s o□□r□a□ □e□ters*

The cities were isolated but arranged into networks.

The Middle East had the densest and most continuous networks throughout history as a result of the large hydraulic civilizations that formed the economic backbone of these networks. Following the phase of the large Mesopotamian cities and the Turkish peninsula, you had the large Arab cities of Baghdad, Damascus, Cairo, Aleppo and their satellite towns, which assured the growth and prosperity of this part of the world.

Central Asia also had very early urban networks, which supported commercial traffic over long roads, valleys and passes. Spilling out of Persian territory (Ispahan), the Persian civilization carried its urban landscapes (mosques with golden, ceramic domes, a central courtyard) as far as Afghanistan (Herat), Uzbekistan (Samarkand, Bukhara) and Iraq (Samara, Najaf).

In the Arabian peninsula and the Sahara, networks were more tenuously maintained, given the vast stretches and the hyper-aridity in these regions. The largest roads were oriented north-south. However, the distances between the cities were much larger here than in the Middle East. The hyper-arid regions between the urban hubs on the northern and southern edges only contained isolated groups: Laayoune, Tamanrasset and Koufra.

This smaller urban dynamic in the Sahara was probably offset by the role of the very active, rural markets, which were sites for exchanges and meetings. Thus, in the Maghrib, rural souks were likely to have substituted intermediaries between the "bled" (villages) and the cities, the result of urban underdevelopment (Troin 1975).

In addition to these cities, born out of a relational life, there were pilgrimage spots. The pilgrim traffic added to, or crossed the path of commercial traffic. There were Mecca and Medina, the holy cities of the Sunnis; Meched, the holy city of the Shias; Jerusalem, the holy city of Christians; and there were also smaller cities, important regionally, such as Chinguetti, Najaf, Karbala and Ghardaïa.

### 3.5.5. *Capitals under the sun*

Desert cities inserted themselves into the landscape along a continuum, going from the villages with a small ksar, to an urban ksar, to a medina, to a capital. Typically, across all levels of communities, we find the same urbanistic configuration, the same maze of streets and impasses, the same desire for bioclimatic and intimist urbanism, with thick walls, snaking covered streets. However, the structures and functions differ from one urban level to another.

A medina differs from a simple ksar by having an external rampart, perhaps reinforced by a ditch. There are gates, a casbah and a large central mosque, neighborhood mosques, a market or open square, stalls set up by artisans and shops. The rich families stay close to the central mosques, and the poorer, more rural households are located close to the external ramparts.

There are all the features that distinguish the Arab-Muslim village, giving it its particular character, which can be seen from Morocco to Pakistan (Figure 3.22).

History has also given deserts the capitals to empires, which were of a wholly different scale in terms of dimensions and populations. Nineveh, Babylon, Cairo, Baghdad, etc., at their peak, were among the most populous cities on Earth. Their landscapes were marked by large monuments, commercial neighborhoods and gardens on the peripheries (ghouta). Aleppo had the largest covered markets (khan) in the world.

To the residents of these cities, as well as foreigners from other lands, these capitals contributed to the creation of a very particular image of deserts.

### 3.5.6. *The city and water*

The city went into the desert. A fortune was needed to supply it with domestic water. This was no small matter when it involved a city with over tens of thousands of residents, or even hundreds of thousands.

Finding urban water has always been an overwhelming preoccupation of the Aediles or leaders of these cities. Archaeologists have revealed the complexity of the solutions developed in the past (Gentelle 2003).

The solutions differed from city to city and would often change across periods. The city of Nineveh was fed by canals 80 km long, and Mari was fed by a canal 120 km long. Petra had a network of cisterns. Each of the large cities in the Punjab had their own diversion dams. Ouargla had a network of

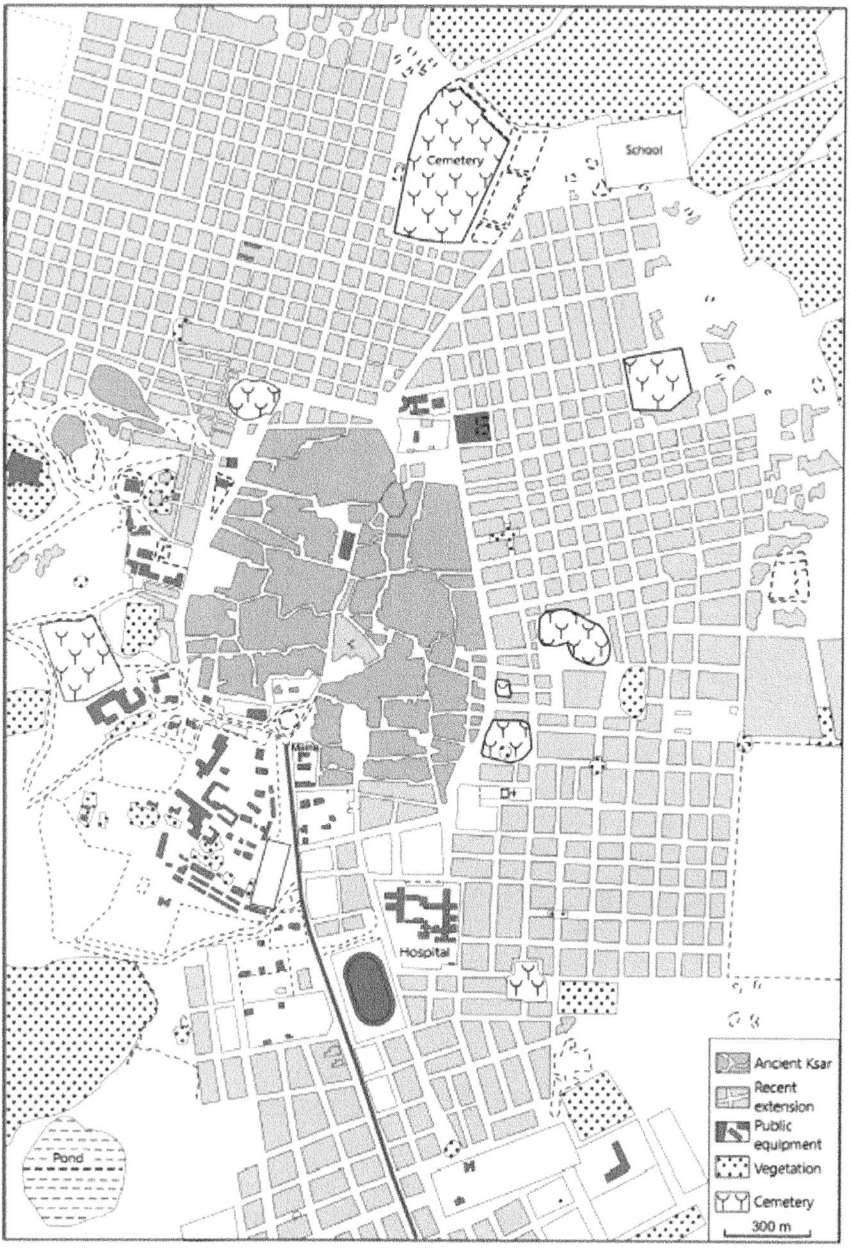


**Figure 3.22.** *A desert city, an Islamic city, Timbuktu. A traditional African city, which was reconstructed as a medina after a fire in the 15th century. An orthogonal extension was added in the 20th century*

well□diggers, responsible for deep, lined wells. Samarkand turned its back on its river but took water from □2 km away. Then there were Baghdad, □almyra, Amman , etc., and each desert city was a unique case.

□odern technology has further enriched this panoply of techniques: Riyadh draws its water from the □ersian □ulf by distilling seawater through giant processing centers and then transporting it across the desert to the capital, through pipes stretching □00 km (□igure 3.23). Then there is Tamanrasset, which has been supplied by a 750 km pipe through the desert for some years now.

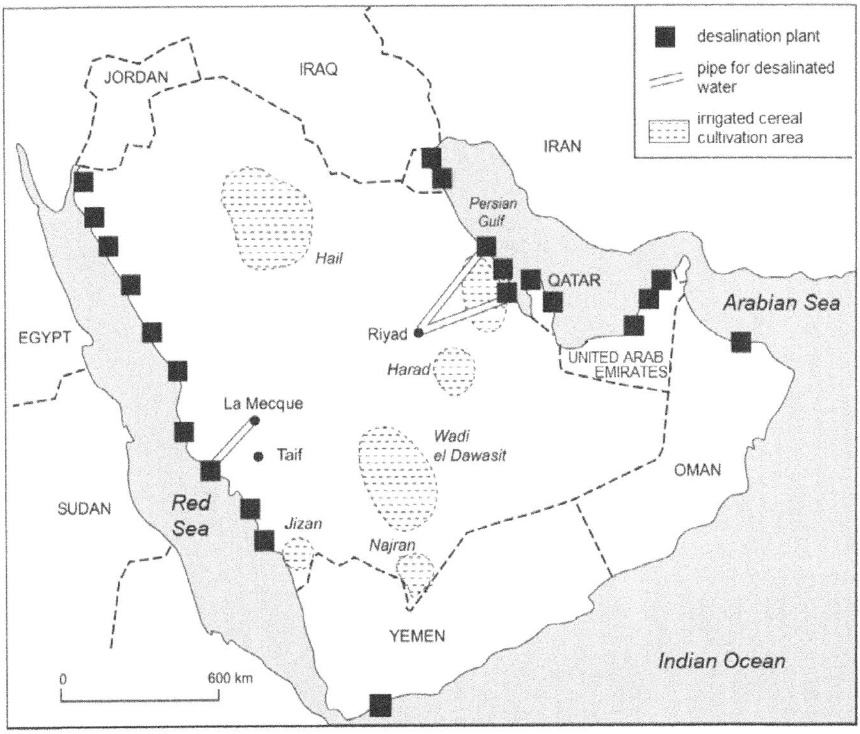


**Figure □.□□.** *Supplying water in the desert. A belt of seawater desalination units across the Arabian peninsula (Mutin 2011)*

### 3.5.7. *The city and the desert today*

Dead cities, forgotten cities and thriving cities make up the urban legacy of the deserts over centuries or millennia.

Today, the most striking feature is the rising urbanization. In a truly spectacular rise, the number of cities has doubled or tripled, their size has increased by three or five times, the rate of urbanization is 60%, or even 80% in many lands across the Arid Diagonal. The oasis-dwellers and nomads are largely outnumbered today by the new inhabitants of the deserts: the urbanites.

A statistical increase reflects the population growth in the deserts at the same time as the economic progress of these lands.

The second feature, which is also striking, is the standardization of these "new" cities. The concept of adapting to a bioclimatic environment, and of adapting to desert culture have been forgotten in a "copy-paste" format that is largely the same around the world. The use of building-blocks, repetitive blocks of buildings, marble and smoked-glass malls are a far cry from the bioclimatic architecture and intimist urbanism. The skylines and urban landscapes of the Gulf countries are inspired more by American urban landscapes than by the cities of the Old World.

The very fundamentals of urban civilization in deserts has been shaken up.

## 3.6. Conclusion

In these vast deserts of the Old World, humans built up a "desert civilization" over centuries, a complex civilization with multiple facets that were intertwined, sometimes antagonistic, but also often complementary.

The dry regions are the only places where two societies – the mobile livestock breeders and the sedentary farmers – with their complementary economies, have cohabited for millennia (Mainguet 2003).

Indeed, we have seen that there are actually three cohabiting societies, and we have tried to demystify this civilization by examining its three elements.

Hydraulic societies laid the foundations for populations in the desert and also laid the basis for an economy through the isolated pockets of the oases, and the large agricultural spots of the irrigated plains.

Bedouin societies, with firm control over their territories, put networks in place, built up over short and long distances, for the constitution and protection of caravans. They exerted pressure on sedentary societies (land taxes) and resisted the neighboring cities (counter-powers).

Desert cities were born of trade: they provided economic hubs. To do this, the cities needed the oases as relay posts and needed the tribes for the logistics of the caravans. Wielding power, they were able to regulate the tensions that arose between the oases and the Bedouins. The urban societies thus combined features of the Bedouin societies and hydraulic societies.

In them, the deserts found their unity.

Today, the particular features of these three elements, sketched here so clearly, have begun to fade away. Urbanization, normalization and modernization have led to deserts becoming annexes of the West.

In the middle of the 2010s, the geopolitical upheavals that rocked many of these territories have changed their fundamentals, blurred the scene and prevented researchers from staying in contact with this part of humanity.

A desert civilization still exists, but to understand it one must be able to read between the lines.

## 3.7. References


Bernus, E. (2006). "La vie des peuples nomades." *Le livre des déserts. Itinéraires scientifiques, litéraires et spirituels*. Ed. by B. Doucey. Paris: R. Laffont, pp. 455–507.

Berque, J. (1958). *Les Villes. Entretiens interdisciplinaires sur les sociétés musulmanes*. 6th section. Paris: École Pratique des Hautes Études.

Bethemont, J. (2000). *Géographie de la Méditerranée*. Collection U. Paris : Armand Colin. 313 p.

Capot-Rey, R. (1953). *Le Sahara français*. Paris : Presses Universitaires de France. 564 p.

Clauzel, J. (1960). *L'exploitation des salines de Taoudéni*. Algiers: Institut de Recherches Sahariennes. 150 pp.

Côte, M. (2014). *Le Sahara, barrière ou pont*. Aix-en-Provence : Presses Universitaires de Provence. 158 p.

Côte, M. (2002). “ Des oasis aux zones de mise en valeur : l’étonnant renouveau de l’agriculture saharienne. Revue géographique des pays méditerranéens”. *Méditerranée* 99 (3-4) : *Le Sahara, “cette autre Méditerranée” (Fernand Braudel).* Edited by M. Côte, p. 5-14. DOI : 10.3406/medit.2002.3253. URL : http ://www.persee.fr/doc/medit_0025-8296_2002_num_99_3_3251.

— (2004). “ Une agriculture pionnière au Fezzan”. *La nouvelle Libye, géopolitique, espaces et sociétés au lendemain de l’embargo.* Edited by O. Pliez. Paris : Karthala-Iremam.

De Planhol, X. and P. Rognon (1970). *Les zones tropicales arides et subtropicales.* Paris: Armand Colin. 487 pp.

Demangeot, J. and E. Bernus (2001). *Les milieux désertiques.* Paris : Armand Colin.

Durand-Dastes, F. and G. Mutin (1995). *Géographie universelle. Afrique du Nord, Monde Indien.* Edited by R. Brunet. 8 vols. Paris : Belin/GIP RECLUS.

Echallier, J.-C. (1968). “Essai sur l’habitat sédentaire traditionnel au Sahara algérien”. Paris : Institut de l’Urbanisme.

Gallais, J. (1967). *Le delta intérieur du Niger, étude de géographie régionale.* 2 vols. Dakar : IFAN. 621 p.

Gentelle, P. (2003). *Traces d’eau: un géographe chez les archéologues.* Paris: Éditions Belin. 239 pp.

Goblot, H. (1979). *Les qanats, une technique d’acquisition de l’eau.* Paris : Mouton. 236 p.

Heathcote, R. (1983). *The Arid Lands, their Use and Abuse.* London: Longman. 323 p.

Hulum, J. (1959). *La géographie de l’Afghanistan. Étude d’un pays aride.* Scandinavian University Books — Publications de l’Institut de Géographie de l’Université d’Aarhus 10. Copenhagen : Gyldendal. 421 p.

Ibn Battuta (1982). *Voyages.* Trans. by C. Defremy and B. Sanguinetti. Paris: François Maspero.

López Austin, A. and L. López-Luján (2014). *El pasado indígena.* 3rd ed. Fideicomiso Historia de las Américas. Mexico: Fondo de Cultura Económica, Colmex, FHA. 320 pp.

Lugan, B. (2008). *Histoire de l’Afrique, des origines à nos jours.* Paris: Ellipses. 1245 pp.

Mainguet, M. (2003). *Les pays secs, environnement et développement.* Paris: Ellipses. 160 pp.

Monod, T. (1989). *Méharées. Explorations au vrai Sahara.* BABEL Terres d’aventure. Arles: Actes Sud. 332 pp.

— (1994). *Thésaurus.* Arles: Actes Sud. 1423 pp.

Petit-Maire, N. (2002). *Sahara - sous les sables... des lacs.* Paris: CNRS. 127 pp.

Planhol, X. de and P. Rognon (1970). *Les zones tropicales arides et subtropicales* Paris : Armand Colin. 487 p.

Puigaudeau, O. du (2001). *Le sel du désert*. Paris: Phébus. 227 pp.

Rousseaux, V. (2004). *L'urbanisation au Maghreb, le langage des cartes*. Aix-en-Provence : Presses Universitaires de Provence. 254 p.

# List of Authors

Yann CALLOT
Lumière University Lyon 2
France

Marc CÔTE
University of Provence
Aix-Marseille I
France

Yvette DEWOLF
Paris Diderot University – Paris 7
France

# Index

## S



## Y, Z

Other titles from

ISTE

in

Ecological Science

## 2021

JOLY Fernand, BOURRIÉ Guilhem
*Mankind and Deserts 2: Water and Salts*

## 2020

JOLY Fernand, BOURRIÉ Guilhem
*Mankind and Deserts 1: Deserts, Aridity, Exploration and Conquests*

BRUSLÉ Jacques, QUIGNARD Jean-Pierre
*Fish Behavior 1: Eco-ethology*
*Fish Behavior 2: Ethophysiology*

LE FLOCH Stéphane
*Remote Detection and Maritime Pollution: Chemical Spill Studies*

MIGON Christophe, NIVAL Paul, SCIANDRA Antoine
*The Mediterranean Sea in the Era of Global Change 1: 30 Years of Multidisciplinary Study of the Ligurian Sea*
*The Mediterranean Sea in the Era of Global Change 2: 30 Years of Multidisciplinary Study of the Ligurian Sea*

ROSSIGNOL Jean-Yves
*Climatic Impact of Activities: Methodological Guide for Analysis and Action*

## 2019

AMIARD Jean-Claude
*Industrial and Medical Nuclear Accidents: Environmental, Ecological, Health and Socio-economic Consequences*
*(Radioactive Risk SET – Volume 2)*
*Nuclear Accidents: Prevention and Management of an Accidental Crisis*
*(Radioactive Risk SET – Volume 3)*

BOULEAU Gabrielle
*Politicization of Ecological Issues: From Environmental Forms to Environmental Motives*

DAVID Valérie
*Statistics in Environmental Sciences*

GIRAULT Yves
*UNESCO Global Geoparks: Tension Between Territorial Development and Heritage Enhancement*

KARA Mohamed Hichem, QUIGNARD Jean-Pierre
*Fishes in Lagoons and Estuaries in the Mediterranean 2: Sedentary Fish*
*Fishes in Lagoons and Estuaries in the Mediterranean 3A: Migratory Fish*
*Fishes in Lagoons and Estuaries in the Mediterranean 3B: Migratory Fish*

OUVRARD Benjamin, STENGER Anne
*Incentives and Environmental Policies: From Theory to Empirical Novelties*

## 2018

AMIARD Jean-Claude
*Military Nuclear Accidents: Environmental, Ecological, Health and Socio-economic Consequences*
*(Radioactive Risk SET – Volume 1)*

FLIPO Fabrice
*The Coming Authoritarian Ecology*

GUILLOUX Bleuenn
*Marine Genetic Resources, R&D and the Law 1: Complex Objects of Use*

KARA Mohamed Hichem, QUIGNARD Jean-Pierre
*Fishes in Lagoons and Estuaries in the Mediterranean 1: Diversity, Bioecology and Exploitation*

## 2016

BAGNÈRES Anne-Geneviève, HOSSAERT-MCKEY Martine
*Chemical Ecology*

## 2014

DE LARMINAT Philippe
*Climate Change: Identification and Projections*